STUDENT SOLUTIONS MANUAL

to accompany

APPLIED CALCULUS

THIRD EDITION

Deborah Hughes-Hallett
University of Arizona
Andrew M. Gleason
Harvard University
Patti Frazer Lock
St. Lawrence University
Daniel E. Flath
Macalester College

et al.

Prepared by:

Elliot J. Marks
Ernie S. Solheid
Ray Cannon

WILEY

John Wiley & Sons, Inc.

Cover Photo: ©Mike Yamashita/Taxi/Getty Images

To order books or for customer service please, call 1-800-CALL WILEY (225-5945).

ISBN-13   978- 0-471-73925-8
ISBN-10   0-471-73925-1

Printed in the United States of America

10 9 8 7 6 5 4 3

Printed and bound by Bind Rite Graphics

CONTENTS

CHAPTER ONE

Solutions for Section 1.1

1. Dan runs 5 kilometers in 23 minutes.

5. If there are no workers, there is no productivity, so the graph goes through the origin. At first, as the number of workers increases, productivity also increases. As a result, the curve goes up initially. At a certain point the curve reaches its highest level, after which it goes downward; in other words, as the number of workers increases beyond that point, productivity decreases. This might, for example, be due either to the inefficiency inherent in large organizations or simply to workers getting in each other's way as too many are crammed on the same line. Many other reasons are possible.

9. Looking at the graph, we see that the point on the graph with an x-coordinate of 5 has a y-coordinate of 2. Thus

$$f(5) = 2.$$

13. The number of species of algae is low when there are few snails or lots of snails. The greatest number of species of algae (about 10) occurs when the number of snails is at a medium level (around 125 snails per square meter.) The graph supports the statement that diversity peaks at intermediate predation levels.

17. (a) From the graph, we see $f(3) = 0.14$. This means that after 3 hours, the level of nicotine is 0.14 mg.
   (b) About 4 hours.
   (c) The vertical intercept is 0.4. It represents the level of nicotine in the blood right after the cigarette is smoked.
   (d) A horizontal intercept would represent the value of t when $N = 0$, or the number of hours until all nicotine is gone from the body.

21. (a) In 1970, fertilizer use in the US was about 13 million tons, in India about 2 million tons, and in the former Soviet Union about 10 million tons.
   (b) Fertilizer use in the US rose steadily between 1950 and 1980 and has stayed relatively constant at about 18 million tons since then. Fertilizer use in India has risen steadily throughout this 50-year period. Fertilizer use in the former Soviet Union rose rapidly between 1950 and 1985 and then declined very rapidly between 1985 and 2000.

Solutions for Section 1.2

1. Rewriting the equation as

$$y = -\frac{12}{7}x + \frac{2}{7}$$

shows that the line has slope $-12/7$ and vertical intercept $2/7$.

5. The slope is $(3 - 2)/(2 - 0) = 1/2$. So the equation of the line is $y = (1/2)x + 2$.

9. The slope is positive for lines l_1 and l_2 and negative for lines l_3 and l_4. The y-intercept is positive for lines l_1 and l_3 and negative for lines l_2 and l_4. Thus, the lines match up as follows:
   (a) l_1
   (b) l_3
   (c) l_2
   (d) l_4

13. This is a linear function with vertical intercept 25 and slope 0.05. The formula for the monthly charge is $C = 25 + 0.05m$.

17. (a) We know that the function for q in terms of p will take on the form

$$q = mp + b.$$

We know that the slope will represent the change in q over the corresponding change in p. Thus

$$m = \text{slope} = \frac{4 - 3}{12 - 15} = \frac{1}{-3} = -\frac{1}{3}.$$

Thus, the function will take on the form

$$q = -\frac{1}{3}p + b.$$

Substituting the values $q = 3, p = 15$, we get

$$3 = -\frac{1}{3}(15) + b$$
$$3 = -5 + b$$
$$b = 8.$$

Thus, the formula for q in terms of p is

$$q = -\frac{1}{3}p + 8.$$

(b) We know that the function for p in terms of q will take on the form

$$p = mq + b.$$

We know that the slope will represent the change in p over the corresponding change in q. Thus

$$m = \text{slope} = \frac{12 - 15}{4 - 3} = -3.$$

Thus, the function will take on the form

$$p = -3q + b.$$

Substituting the values $q = 3, p = 15$ again, we get

$$15 = (-3)(3) + b$$
$$15 = -9 + b$$
$$b = 24.$$

Thus, a formula for p in terms of q is

$$p = -3q + 24.$$

21. (a) This could be a linear function because w increases by 5 as h increases by 1.
(b) We find the slope m and the intercept b in the linear equation $w = b + mh$. We first find the slope m using the first two points in the table. Since we want w to be a function of h, we take

$$m = \frac{\Delta w}{\Delta h} = \frac{171 - 166}{69 - 68} = 5.$$

Substituting the first point and the slope $m = 5$ into the linear equation $w = b + mh$, we have $166 = b + (5)(68)$, so $b = -174$. The linear function is

$$w = 5h - 174.$$

The slope, $m = 5$, is in units of pounds per inch.
(c) We find the slope and intercept in the linear function $h = b + mw$ using $m = \Delta h / \Delta w$ to obtain the linear function

$$h = 0.2w + 34.8.$$

Alternatively, we could solve the linear equation found in part (b) for h. The slope, $m = 0.2$, has units inches per pound.

25. (a) We find the slope m and intercept b in the linear equation $S = b + mt$. To find the slope m, we use

$$m = \frac{\Delta S}{\Delta t} = \frac{66 - 113}{50 - 0} = -0.94.$$

When $t = 0$, we have $S = 113$, so the intercept b is 113. The linear formula is

$$S = 113 - 0.94t.$$

(b) We use the formula $S = 113 - 0.94t$. When $S = 20$, we have $20 = 113 - 0.94t$ and so $t = 98.9$. If this linear model were correct, the average male sperm count would drop below the fertility level during the year 2038.

29. (a) If old and new formulas give the same MHR for females, we have

$$226 - a = 208 - 0.7a,$$

so

$$a = 60 \text{ years.}$$

For males we need to solve the equation

$$220 - a = 208 - 0.7a,$$

so

$$a = 40 \text{ years.}$$

(b) The old formula starts at either 226 or 220 and decreases. The new formula starts at 208 and decreases slower than the old formula. Thus, the new formula predicts a lower MHR for young people and a higher MHR for older people, so the answer is (ii).

(c) Under the old formula,

$$\text{Heart rate reached} = 0.85(220 - 65) = 131.75 \text{ beats/minute,}$$

whereas under the new formula,

$$\text{Heart rate reached} = 0.85(208 - 0.7 \cdot 65) = 138.125 \text{ beats/minute.}$$

The difference is $138.125 - 131.75 = 6.375$ beats/minute more under the new formula.

Solutions for Section 1.3

1. The graph shows a concave down function.

5. As t increases w decreases, so the function is decreasing. The rate at which w is decreasing is itself decreasing: as t goes from 0 to 4, w decreases by 42, but as t goes from 4 to 8, w decreases by 36. Thus, the function is concave up.

9. The average rate of change R between $x = 1$ and $x = 3$ is

$$
\begin{aligned}
R &= \frac{f(3) - f(1)}{3 - 1} \\
&= \frac{18 - 2}{2} \\
&= \frac{16}{2} \\
&= 8.
\end{aligned}
$$

13. (a) This function is increasing and the graph is concave down.

(b) From the graph, we estimate that when $t = 5$, we have $L = 70$ and when $t = 15$, we have $L = 130$. Thus,

$$\text{Average rate of change of length} = \frac{\Delta L}{\Delta t} = \frac{130 - 70}{15 - 5} = 6 \text{ cm/year.}$$

During this ten year period, the sturgeon's length increased at an average rate of 6 centimeters per year.

17. (a) Between 1999 and 2004, we have

$$
\begin{aligned}
\text{Change in sales} &= \text{Sales in 2004} - \text{Sales in 1999} \\
&= 29{,}261 - 20{,}367 \\
&= 8894 \text{ million dollars.}
\end{aligned}
$$

(b) Over the same period, we have

$$\begin{aligned}
\begin{array}{c}\text{Average rate of change in sales}\\\text{between 1999 and 2004}\end{array} &= \frac{\text{Change in sales}}{\text{Change in time}}\\[2mm]
&= \frac{\text{Sales in 2004} - \text{Sales in 1999}}{2004 - 1999}\\[2mm]
&= \frac{29{,}261 - 20{,}367}{2004 - 1999}\\[2mm]
&= \frac{8894}{5}\\[2mm]
&= 1778.8 \text{ million dollars per year.}
\end{aligned}$$

This means that Pepsico's sales increased on average by 1778.8 million dollars per year between 1999 and 2004.

21. (a) (i) Between the 6th and 8th minutes the concentration changes from 0.336 to 0.298 so we have

$$\text{Average rate of change} = \frac{0.298 - 0.336}{8 - 6} = -0.019 \text{ (mg/ml)/min.}$$

(ii) Between the 8th and 10th minutes we have

$$\text{Average rate of change} = \frac{0.266 - 0.298}{10 - 8} = -0.016 \text{ (mg/ml)/min.}$$

(b) The signs are negative because the concentration is decreasing. The magnitude of the decrease is decreasing because, as the concentration falls, the rate at which it falls decreases.

25. Each of these questions can also be answered by considering the slope of the line joining the two relevant points.

(a) The average rate of change is positive if the volume of water is increasing with time and negative if the volume of water is decreasing.
  (i) Since volume is rising from 500 to 1000 from $t = 0$ to $t = 5$, the average rate of change is positive.
  (ii) We can see that the volume at $t = 10$ is greater than the volume at $t = 0$. Thus, the average rate of change is positive.
  (iii) We can see that the volume at $t = 15$ is lower than the volume at $t = 0$. Thus, the average rate of change is negative.
  (iv) We can see that the volume at $t = 20$ is greater than the volume at $t = 0$. Thus, the average rate of change is positive.

(b) (i) The secant line between $t = 0$ and $t = 5$ is steeper than the secant line between $t = 0$ and $t = 10$, so the slope of the secant line is greater on $0 \le t \le 5$. Since average rate of change is represented graphically by the slope of a secant line, the rate of change in the interval $0 \le t \le 5$ is greater than that in the interval $0 \le t \le 10$.
  (ii) The slope of the secant line between $t = 0$ and $t = 20$ is greater than the slope of the secant line between $t = 0$ and $t = 10$, so the rate of change is larger for $0 \le t \le 20$.

(c) The average rate of change in the interval $0 \le t \le 10$ is about

$$\frac{750 - 500}{10} = \frac{250}{10} = 25 \quad \text{cubic meters per week}$$

This tells us that for the first ten weeks, the volume of water is growing at an average rate of 25 cubic meters per week.

29. Between 0 and 21 years,

$$\text{Average rate of change} = -\frac{9}{21} = -0.429 \text{ beats per minute per year,}$$

whereas between 0 and 33 years,

$$\text{Average rate of change} = -\frac{26}{33} = -0.788 \text{ beats per minute per year.}$$

Because the average rate of change is negative, a decreasing function is suggested. Also, as age increases, the average rate of change decreases, suggesting the graph of the function is concave down. (Since the average rate of change is negative and increasing in absolute value, this rate is decreasing.)

Solutions for Section 1.4

1. (a) Since cost is less than revenue for quantities in the table between 20 and 60 units, production appears to be profitable between these values.

   (b) Profit = Revenue − Cost is show in Table 1.1. The maximum profit is obtained at a production level of about 40 units.

Table 1.1

Quantity	0	10	20	30	40	50	60	70	80
Cost ($)	120	400	600	780	1000	1320	1800	2500	3400
Revenue ($)	0	300	600	900	1200	1500	1800	2100	2400
Profit	−120	−100	0	120	200	180	0	−400	−1000

5. (a) The fixed costs are the price of producing zero units, or $C(0)$, which is the vertical intercept. Thus, the fixed costs are roughly $75. The variable cost is the slope of the line. We know that

$$C(0) = 75$$

and looking at the graph we can also tell that

$$C(30) = 300.$$

Thus, the slope or the variable cost is

$$\text{Variable cost} = \frac{300 - 75}{30 - 0} = \frac{225}{30} = 7.50 \text{ dollars per unit.}$$

   (b) Looking at the graph it seems that

$$C(10) \approx 150.$$

Alternatively, using what we know from parts (a) and (b) we know that the cost function is

$$C(q) = 7.5q + 75.$$

Thus,

$$C(10) = 7.5(10) + 75 = 75 + 75 = \$150.$$

The total cost of producing 10 items is $150.

9. (a) We know that the function for the cost of running the theater is of the form

$$C = mq + b$$

where q is the number of customers, m is the variable cost and b is the fixed cost. Thus, the function for the cost is

$$C = 2q + 5000.$$

We know that the revenue function is of the form

$$R = pq$$

where p is the price charged per customer. Thus, the revenue function is

$$R = 7q.$$

The theater makes a profit when the revenue is greater than the cost, that is when

$$R > C.$$

Substituting $R = 7q$ and $C = 2q + 5000$, we get

$$R > C$$
$$7q > 2q + 5000$$
$$5q > 5000$$
$$q > 1000.$$

Thus, the theater makes a profit when it has more than 1000 customers.

(b) The graph of the two functions is shown in Figure 1.1.

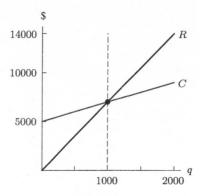

Figure 1.1

13. **(a)** A company with little or no fixed costs would be one that does not need much start-up capital and whose costs are mainly on a per unit basis. An example of such a company is a consulting company, whose major expense is the time of its consultants. Such a company would have little fixed costs to worry about.

   (b) A company with little or no variable costs would be one that can produce a product with little or no additional costs per unit. An example is a computer software company. The major expense of such a company is software development, a fixed cost. Additional copies of its software can be very easily made. Thus, its variable costs are rather small.

17. **(a)** The statement $f(12) = 60$ says that when $p = 12$, we have $q = 60$. When the price is \$12, we expect to sell 60 units.

   (b) Decreasing, because as price increases, we expect less to be sold.

21. **(a)** We know that the cost function will be of the form

$$C = mq + b$$

where m is the variable cost and b is the fixed cost. In this case this gives

$$C = 5q + 7000.$$

We know that the revenue function is of the form

$$R = pq$$

where p is the price per shirt. Thus in this case we have

$$R = 12q.$$

   (b) We are given

$$q = 2000 - 40p.$$

We are asked to find the demand when the price is \$12. Plugging in $p = 12$ we get

$$q = 2000 - 40(12) = 2000 - 480 = 1520.$$

Given this demand we know that the cost of producing $q = 1520$ shirts is

$$C = 5(1520) + 7000 = 7600 + 7000 = \$14,600.$$

The revenue from selling $q = 1520$ shirts is

$$R = 12(1520) = \$18,240.$$

Thus the profit is

$$\pi(12) = R - C$$

or in other words

$$\pi(12) = 18,240 - 14,600 = \$3640.$$

   (c) Since we know that

$$q = 2000 - 40p,$$
$$C = 5q + 7000,$$

and
$$R = pq,$$
we can write
$$C = 5q + 7000 = 5(2000 - 40p) + 7000 = 10{,}000 - 200p + 7000 = 17{,}000 - 200p$$
and
$$R = pq = p(2000 - 40p) = 2000p - 40p^2.$$
We also know that the profit is the difference between the revenue and the cost so
$$\pi(p) = R - C = 2000p - 40p^2 - (17{,}000 - 200p) = -40p^2 + 2200p - 17{,}000.$$

(d) Looking at Figure 1.2 we see that the maximum profit occurs when the company charges about \$27.50 per shirt. At this price, the profit is about \$13,250.

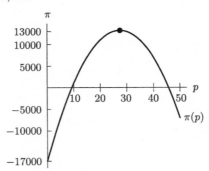

Figure 1.2

25. (a) If we think of q as a linear function of p, then q is the dependent variable, p is the independent variable, and the slope $m = \Delta q / \Delta p$. We can use any two points to find the slope. If we use the first two points, we get
$$\text{Slope} = m = \frac{\Delta q}{\Delta p} = \frac{460 - 500}{18 - 16} = \frac{-40}{2} = -20.$$

The units are the units of q over the units of p, or tons per dollar. The slope tells us that, for every dollar increase in price, the number of tons sold every month will decrease by 20.

To write q as a linear function of p, we need to find the vertical intercept, b. Since q is a linear function of p, we have $q = b + mp$. We know that $m = -20$ and we can use any of the points in the table, such as $p = 16$, $q = 500$, to find b. Substituting gives
$$q = b + mp$$
$$500 = b + (-20)(16)$$
$$500 = b - 320$$
$$820 = b.$$

Therefore, the vertical intercept is 820 and the equation of the line is
$$q = 820 - 20p.$$

(b) If we now consider p as a linear function of q, we have
$$\text{Slope} = m = \frac{\Delta p}{\Delta q} = \frac{18 - 16}{460 - 500} = \frac{2}{-40} = -\frac{1}{20} = -0.05.$$

The units of the slope are dollars per ton. The slope tells us that, if we want to sell one more ton of the product every month, we should reduce the price by \$0.05.

Since p is a linear function of q, we have $p = b + mq$ and $m = -0.05$. To find b, we substitute any point from the table such as $p = 16$, $q = 500$ into this equation:
$$p = b + mq$$
$$16 = b + (-0.05)(500)$$
$$16 = b - 25$$
$$41 = b.$$

The equation of the line is

$$p = 41 - 0.05q.$$

Alternatively, notice that we could have taken our answer to part (a), that is $q = 820 - 20p$, and solved for p.

29. (a) The amount spent on books will be

$$\text{Amount for books} = \$40 \cdot b$$

where b is the number of books bought. The amount of money spent on outings is

$$\text{money spent on outings} = \$10 \cdot s$$

where s is the number of social outings. Since we want to spend all of the \$1000 budget we end up with

$$40b + 10s = 1000.$$

(b)

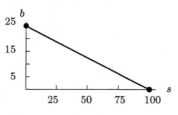

Figure 1.3

(c) Looking at Figure 1.3 we see that the vertical intercept occurs at the point

$$(0, 25)$$

and the horizontal intercept occurs at

$$(100, 0).$$

The vertical intercept tells us how many books we would be able to buy if we wanted to spend all of the budget on books. That is, we could buy at most 25 books. The horizontal intercept tells how many social outings we could afford if we wanted to spend all of the budget on outings. That is, we would be able to go on at most 100 outings.

33. The original supply equation, $q = 4p - 20$, tells us that

$$\text{Quantity supplied} = 4 \left(\begin{array}{c} \text{Amount per unit} \\ \text{received by suppliers} \end{array} \right) - 20.$$

The suppliers receive only $p - 2$ dollars per unit because \$2 goes to the government as taxes. Thus, the new supply equation is

$$q = 4(p - 2) - 20 = 4p - 28.$$

See Figure 1.4.

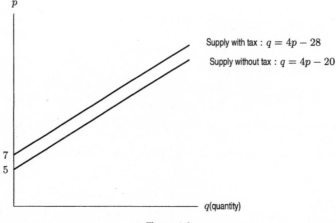

Figure 1.4

37. (a) The original supply equation, $q = 3p - 50$, tells us that

$$\text{Quantity supplied } = 3 \left(\begin{array}{c} \text{Amount received per} \\ \text{unit by suppliers} \end{array} \right) - 50.$$

The suppliers receive only $p - 0.05p = 0.95p$ dollars per unit because 5% of the price goes to the government. Thus, the new supply equation is

$$q = 3(0.95p) - 50 = 2.85p - 50.$$

The demand equation, $q = 100 - 2p$ is unchanged.

(b) At the equilibrium, supply equals demand:

$$2.85p - 50 = 100 - 2p$$
$$4.85p = 150$$
$$p = \$30.93.$$

The equilibrium price is \$30.93. The equilibrium quantity q is the quantity demanded at the equilibrium price:

$$q = 100 - 2(30.93) = 38.14 \text{ units.}$$

(c) Since the pretax price was \$30 and consumers' new price is \$30.93,

$$\text{Tax paid by consumers} = 30.93 - 30 = \$0.93$$

The supplier keeps $0.95p = 0.95(30.93) = \$29.38$ per unit, so

$$\text{Tax paid by suppliers} = 30 - 29.38 = \$0.62.$$

The total tax per unit paid by consumers and suppliers together is $0.93 + 0.62 = \$1.55$ per unit.

(d) The government receives \$1.55 per unit on 38.14 units. The total tax collected is $(1.55)(38.14) = \$59.12$.

Solutions for Section 1.5

1. (a) Town (i) has the largest percent growth rate, at 12%.

(b) Town (ii) has the largest initial population, at 1000.

(c) Yes, town (iv) is decreasing in size, since the decay factor is 0.9, which is less than 1.

5. (a) Since the price is decreasing at a constant absolute rate, the price of the product is a linear function of t. In t days, the product will cost $80 - 4t$ dollars.

(b) Since the price is decreasing at a constant relative rate, the price of the product is an exponential function of t. In t days, the product will cost $80(0.95)^t$.

9. (a) We see from the base 1.0126 that the percent rate of growth is 1.26% per year.

(b) The initial population (in the year 2004) is 6.4 billion people. When $t = 6$, we have $P = 6.4(1.0126)^5 = 6.90$. The predicted world population in the year 2010 is 6.90 billion people.

(c) We have

$$\text{Average rate of change } = \frac{\Delta P}{\Delta t} = \frac{6.90 - 6.4}{2010 - 2004} = 0.083 \text{ billion people per year.}$$

The population of the world is increasing by about 83 million people per year.

13. This looks like an exponential function $y = Ca^t$. The y-intercept is 500 so we have $y = 500a^t$. We use the point $(3, 2000)$ to find a:

$$y = 500a^t$$
$$2000 = 500a^3$$
$$4 = a^3$$
$$a = 4^{1/3} = 1.59.$$

The formula is $y = 500(1.59)^t$.

17. (a) A linear function must change by exactly the same amount whenever x changes by some fixed quantity. While $h(x)$ decreases by 3 whenever x increases by 1, $f(x)$ and $g(x)$ fail this test, since both change by different amounts between $x = -2$ and $x = -1$ and between $x = -1$ and $x = 0$. So the only possible linear function is $h(x)$, so it will be given by a formula of the type: $h(x) = mx + b$. As noted, $m = -3$. Since the y-intercept of h is 31, the formula for $h(x)$ is $h(x) = 31 - 3x$.

(b) An exponential function must grow by exactly the same factor whenever x changes by some fixed quantity. Here, $g(x)$ increases by a factor of 1.5 whenever x increases by 1. Since the y-intercept of $g(x)$ is 36, $g(x)$ has the formula $g(x) = 36(1.5)^x$. The other two functions are not exponential; $h(x)$ is not because it is a linear function, and $f(x)$ is not because it both increases and decreases.

21. Direct calculation reveals that each 1000 foot increase in altitude results in a longer takeoff roll by a factor of about 1.096. Since the value of d when $h = 0$ (sea level) is $d = 670$, we are led to the formula

$$d = 670(1.096)^{h/1000},$$

where d is the takeoff roll, in feet, and h is the airport's elevation, in feet.

Alternatively, we can write

$$d = d_0 a^h,$$

where d_0 is the sea level value of d, $d_0 = 670$. In addition, when $h = 1000$, $d = 734$, so

$$734 = 670a^{1000}.$$

Solving for a gives

$$a = \left(\frac{734}{670}\right)^{1/1000} = 1.00009124,$$

so

$$d = 670(1.00009124)^h.$$

25. (a) We have

$$\text{Reduced size} = (0.80) \cdot \text{Original size}$$

or

$$\text{Original size} = \frac{1}{(0.80)} \text{Reduced size} = (1.25) \text{Reduced size},$$

so the copy must be enlarged by a factor of 1.25, which means it is enlarged to 125% of the reduced size.

(b) If a page is copied n times, then

$$\text{New size} = (0.80)^n \cdot \text{Original}.$$

We want to solve for n so that

$$(0.80)^n = 0.15.$$

By trial and error, we find $(0.80)^8 = 0.168$ and $(0.80)^9 = 0.134$. So the page needs to be copied 9 times.

Solutions for Section 1.6

1. Taking natural logs of both sides we get

$$\ln(5^t) = \ln 7.$$

This gives

$$t \ln 5 = \ln 7$$

or in other words

$$t = \frac{\ln 7}{\ln 5} \approx 1.209.$$

5. Dividing both sides by 10 we get

$$5 = 3^t.$$

Taking natural logs of both sides gives

$$\ln(3^t) = \ln 5.$$

This gives

$$t \ln 3 = \ln 5$$

or in other words

$$t = \frac{\ln 5}{\ln 3} \approx 1.465.$$

9. Dividing both sides by 2 we get

$$2.5 = e^t.$$

Taking the natural log of both sides gives

$$\ln(e^t) = \ln 2.5.$$

This gives

$$t = \ln 2.5 \approx 0.9163.$$

13. Dividing both sides by P we get

$$\frac{B}{P} = e^{rt}.$$

Taking the natural log of both sides gives

$$\ln(e^{rt}) = \ln\left(\frac{B}{P}\right).$$

This gives

$$rt = \ln\left(\frac{B}{P}\right) = \ln B - \ln P.$$

Dividing by r gives

$$t = \frac{\ln B - \ln P}{r}.$$

17. Initial quantity $= 5$; growth rate $= 0.07 = 7\%$.

21. **(a)** (i) $P = 1000(1.05)^t$; (ii) $P = 1000e^{0.05t}$
    (b) (i) 1629; (ii) 1649

25. $P = P_0(e^{0.2})^t = P_0(1.2214)^t$. Exponential growth because $0.2 > 0$ or $1.2214 > 1$.

29. We want $0.9^t = e^{kt}$ so $0.9 = e^k$ and $k = \ln 0.9 = -0.1054$. Thus $P = 174e^{-0.1054t}$.

33. **(a)** The continuous percent growth rate is 6%.
    (b) We want $P_0 a^t = 100e^{0.06t}$, so we have $P_0 = 100$ and $a = e^{0.06} = 1.0618$. The corresponding function is

$$P = 100(1.0618)^t.$$

The annual percent growth rate is 6.18%. An annual rate of 6.18% is equivalent to a continuous rate of 6%.

37. We find a with $a^t = e^{0.08t}$. Thus, $a = e^{0.08} = 1.0833$. The corresponding annual percent growth rate is 8.33%.

41. Let $t = $ number of years since 1980. Then the number of vehicles, V, in millions, at time t is given by

$$V = 170(1.04)^t$$

and the number of people, P, in millions, at time t is given by

$$P = 227(1.01)^t.$$

There is an average of one vehicle per person when $\frac{V}{P} = 1$, or $V = P$. Thus, we must solve for t the equation:

$$170(1.04)^t = 227(1.01)^t,$$

which implies

$$\left(\frac{1.04}{1.01}\right)^t = \frac{(1.04)^t}{(1.01)^t} = \frac{227}{170}.$$

Taking logs on both sides,

$$t\ln\frac{1.04}{1.01} = \ln\frac{227}{170}.$$

Therefore,

$$t = \frac{\ln\left(\frac{227}{170}\right)}{\ln\left(\frac{1.04}{1.01}\right)} \approx 9.9 \text{ years.}$$

So there was, according to this model, about one vehicle per person in 1990.

Solutions for Section 1.7

1. Every 2 hours, the amount of nicotine remaining is reduced by $1/2$. Thus, after 4 hours, the amount is $1/2$ of the amount present after 2 hours.

Table 1.2

t (hours)	0	2	4	6	8	10
Nicotine (mg)	0.4	0.2	0.1	0.05	0.025	0.0125

From the table it appears that it will take just over 6 hours for the amount of nicotine to reduce to 0.04 mg.

5. (a) We know that the formula for the account balance at time t in an account compounded continuously is given by the formula

$$P(t) = P_0 e^{rt}$$

where P_0 is the initial deposit and r is the annual rate. Thus, in our case the formula would be

$$P(t) = 5000 e^{0.04t}.$$

Substituting the value $t = 8$ we get

$$
\begin{aligned}
P(8) &= 5000 e^{0.04(8)} \\
&= 5000 e^{0.32} \\
&\approx 5000(1.377128) \\
&= 6885.64.
\end{aligned}
$$

Thus, the balance at the end of eight years would be about \$6885.64.

(b) We are asked to solve for the rate, r, that would give us an \$8000 balance at the end of eight years. In other words we are asked to solve for r in the equation

$$8000 = 5000 e^{8r}.$$

Solving we get

$$
\begin{aligned}
8000 &= 5000 e^{8r} \\
e^{8r} &= \frac{8000}{5000} = 1.6 \\
\ln e^{8r} &= \ln 1.6 \\
8r &= \ln 1.6 \\
r &= \frac{\ln 1.6}{8} \approx 0.059.
\end{aligned}
$$

Thus, the rate we would need in order to have a balance of \$8000 at the end of eight years is about 5.9%.

9. (a) Since $P(t)$ is an exponential function, it will be of the form $P(t) = P_0 a^t$, where P_0 is the initial population and a is the base. $P_0 = 200$, and a 5% growth rate means $a = 1.05$. Thus, we get

$$P(t) = 200(1.05)^t.$$

(b) The graph is shown in Figure 1.5.

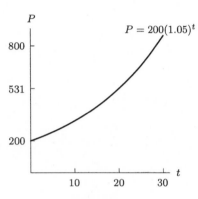

Figure 1.5

(c) Evaluating gives that $P(10) = 200(1.05)^{10} \approx 326$.

(d) From the graph we see that the population is 400 at about $t = 15$, so the doubling time appears to be about 15 years.

13. (a) We want to find t such that

$$0.15Q_0 = Q_0 e^{-0.000121t},$$

so $0.15 = e^{-0.000121t}$, meaning that $\ln 0.15 = -0.000121t$, or $t = \dfrac{\ln 0.15}{-0.000121} \approx 15{,}678.7$ years.

(b) Let T be the half-life of carbon-14. Then

$$0.5Q_0 = Q_0 e^{-0.000121T},$$

so $0.5 = e^{-0.000121T}$, or $T = \dfrac{\ln 0.5}{-0.000121} \approx 5728.5$ years.

17. In both cases the initial deposit was $20. Compounding continuously earns more interest than compounding annually at the same interest rate. Therefore, curve A corresponds to the account which compounds interest continuously and curve B corresponds to the account which compounds interest annually. We know that this is the case because curve A is higher than curve B over the interval, implying that bank account A is growing faster, and thus is earning more money over the same time period.

21. (a) If the CD pays 9% interest over a 10-year period, then $r = 0.09$ and $t = 10$. We must find the initial amount P_0 if the balance after 10 years is $P = \$12{,}000$. Since the compounding is annual, we use

$$P = P_0(1 + r)^t$$
$$12{,}000 = P_0(1.09)^{10},$$

and solve for P_0:

$$P_0 = \frac{12{,}000}{(1.09)^{10}} \approx \frac{12{,}000}{2.36736} = 5068.93.$$

The initial deposit must be $5068.93 if interest is compounded annually.

(b) On the other hand, if the CD interest is compounded continuously, we have

$$P = P_0 e^{rt}$$
$$12{,}000 = P_0 e^{(0.09)(10)}.$$

Solving for P_0 gives

$$P_0 = \frac{12{,}000}{e^{(0.09)(10)}} = \frac{12{,}000}{e^{0.9}} \approx \frac{12{,}000}{2.45960} = 4878.84.$$

The initial deposit must be $4878.84 if the interest is compounded continuously. Notice that to achieve the same result, continuous compounding requires a smaller initial investment than annual compounding. This is to be expected, since continuous compounding yields more money.

25. (a) Using the Rule of 70, we get that the doubling time of the investment is

$$\frac{70}{8} = 8.75.$$

That is, it would take about 8.75 years for the investment to double.

(b) We know that the formula for the balance at the end of t years is

$$B(t) = Pa^t$$

where P is the initial investment and a is 1+(interest per year). We are asked to solve for the doubling time, which amounts to asking for the time t at which

$$B(t) = 2P.$$

Solving, we get

$$2P = P(1 + 0.08)^t$$
$$2 = 1.08^t$$
$$\ln 2 = \ln 1.08^t$$
$$t \ln 1.08 = \ln 2$$
$$t = \frac{\ln 2}{\ln 1.08} \approx 9.01.$$

Thus, the actual doubling time is 9.01 years. And so our estimation by the "rule of 70" was off by a quarter of a year.

29. We assume exponential decay and solve for k using the half-life:

$$e^{-k(5730)} = 0.5 \quad \text{so} \quad k = 1.21 \cdot 10^{-4}.$$

Now find t, the age of the painting:

$$e^{-1.21 \cdot 10^{-4} t} = 0.995, \quad \text{so} \quad t = \frac{\ln 0.995}{-1.21 \cdot 10^{-4}} = 41.43 \text{ years.}$$

Since Vermeer died in 1675, the painting is a fake.

33. In effect, your friend is offering to give you \$17,000 now in return for the \$19,000 lottery payment one year from now. Since $19,000/17,000 = 1.11764\cdots$, your friend is charging you 11.7% interest per year, compounded annually. You can expect to get more by taking out a loan as long as the interest rate is less than 11.7%. In particular, if you take out a loan, you have the first lottery check of \$19,000 plus the amount you can borrow to be paid back by a single payment of \$19,000 at the end of the year. At 8.25% interest, compounded annually, the present value of 19,000 one year from now is $19,000/(1.0825) = 17,551.96$. Therefore the amount you can borrow is the total of the first lottery payment and the loan amount, that is, $19,000 + 17,551.96 = 36,551.96$. So you do better by taking out a one-year loan at 8.25% per year, compounded annually, than by accepting your friend's offer.

Solutions for Section 1.8

1. (a) $g(2 + h) = (2 + h)^2 + 2(2 + h) + 3 = 4 + 4h + h^2 + 4 + 2h + 3 = h^2 + 6h + 11.$
 (b) $g(2) = 2^2 + 2(2) + 3 = 4 + 4 + 3 = 11$, which agrees with what we get by substituting $h = 0$ into (a).
 (c) $g(2 + h) - g(2) = (h^2 + 6h + 11) - (11) = h^2 + 6h.$

5. (a) $f(g(1)) = f(1^2) = f(1) = e^1 = e$
 (b) $g(f(1)) = g(e^1) = g(e) = e^2$
 (c) $f(g(x)) = f(x^2) = e^{x^2}$
 (d) $g(f(x)) = g(e^x) = (e^x)^2 = e^{2x}$
 (e) $f(t)g(t) = e^t t^2$

9. (a) We have $f(g(x)) = f(5x^2) = 2(5x^2) + 3 = 10x^2 + 3.$
 (b) We have $g(f(x)) = g(2x + 3) = 5(2x + 3)^2 = 5(4x^2 + 12x + 9) = 20x^2 + 60x + 45.$
 (c) We have $f(f(x)) = f(2x + 3) = 2(2x + 3) + 3 = 4x + 9.$

13. This graph is the graph of $m(t)$ shifted upward by two units. See Figure 1.6.

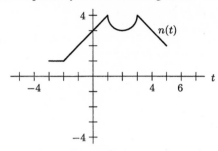

Figure 1.6

17. See Figure 1.7.

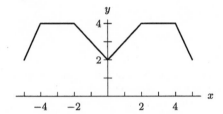

Figure 1.7: Graph of $y = f(x) + 2$

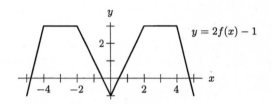

Figure 1.8

21. See Figure 1.8.

25.

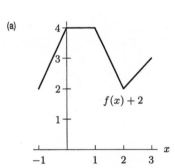

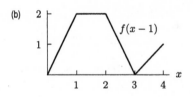

29. For $f(x) + 5$, the graph is shifted 5 upward. See Figure 1.9.

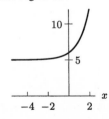

Figure 1.9

33. $g(f(2)) \approx g(0.4) \approx 1.1.$

37. (a) First note that C is negative, because if $C = 0$ then the quantity of pollution $Q = f(t) = A + Be^{Ct}$ does not change over time and if $C > 0$ then $Q = f(t)$ approaches $+\infty$ or $-\infty$ as t approaches ∞, which is not realistic. Since $C < 0$, the quantity $Q = f(t)$ approaches $A + B(0) = A$ as t increases. Since Q can not be negative or 0, we have $A > 0$. Finally, since $Q = f(t)$ is a decreasing function we have $B > 0$.
  (b) Initially we have $Q = f(0) = A + Be^{C(0)} = A + B$.
  (c) We saw in the solution to part (a) that $Q = f(t)$ approaches A as t approaches ∞, so the legal limit, which is the goal of the clean-up project, is A.

Solutions for Section 1.9

1. $y = 5x^{1/2}; k = 5, p = 1/2.$

5. $y = 9x^{10}; k = 9, p = 10.$

9. $y = 8x^{-1}; k = 8, p = -1.$

13. For some constant k, we have $S = kh^2$.

17. The specific heat is larger when the atomic weight is smaller, so we test to see if the two are inversely proportional to each other by calculating their product. This method works because if y is inversely proportional to x, then

$$y = \frac{k}{x} \quad \text{for some constant } k, \quad \text{so} \quad yx = k.$$

Table 1.3 shows that the product is approximately 6 in each case, so we conjecture that $sw \approx 6$, or $s \approx 6/w$.

Table 1.3

Element	Lithium	Magnesium	Aluminium	Iron	Silver	Lead	Mercury
w	6.9	24.3	27.0	55.8	107.9	207.2	200.6
s (cal/deg-gm)	0.92	0.25	0.21	0.11	0.056	0.031	.033
sw	6.3	6.1	5.7	6.1	6.0	6.4	6.6

21. Since

$$S = kM^{2/3}$$

we have

$$18{,}600 = k(70^{2/3})$$

and so

$$k = 1095.$$

We have $S = 1095M^{2/3}$. If $M = 60$, then

$$S = 1095(60^{2/3}) = 16{,}782 \text{ cm}^2.$$

25. Substituting $w = 65$ and $h = 160$, we have
  (a)
$$s = 0.01(65^{0.25})(160^{0.75}) = 1.3 \text{ m}^2.$$

  (b) We substitute $s = 1.5$ and $h = 180$ and solve for w:

$$1.5 = 0.01w^{0.25}(180^{0.75}).$$

We have

$$w^{0.25} = \frac{1.5}{0.01(180^{0.75})} = 3.05.$$

Since $w^{0.25} = w^{1/4}$, we take the fourth power of both sides, giving

$$w = 86.8 \text{ kg}.$$

(c) We substitute $w = 70$ and solve for h in terms of s:

$$s = 0.01(70^{0.25})h^{0.75},$$

so

$$h^{0.75} = \frac{s}{0.01(70^{0.25})}.$$

Since $h^{0.75} = h^{3/4}$, we take the 4/3 power of each side, giving

$$h = \left(\frac{s}{0.01(70^{0.25})} \right)^{4/3} = \frac{s^{4/3}}{(0.01^{4/3})(70^{1/3})}$$

so

$$h = 112.6s^{4/3}.$$

29. (a) The degree of $x^2 + 10x - 5$ is 2 and its leading coefficient is positive.
  (b) In a large window, the function looks like x^2. See Figure 1.10.

Figure 1.10

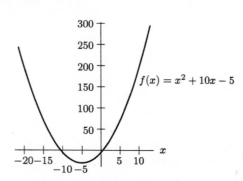

Figure 1.11

  (c) The function has one turning point which is what we expect of a quadratic. See Figure 1.11.

33. (a) The function $f(x) = 100 + 5x - 12x^2 + 3x^3 - x^4$ has degree 4 and a negative leading coefficient.
  (b) In a large window, the function looks like $-x^4$. See Figure 1.12.

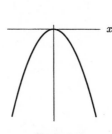

Figure 1.12

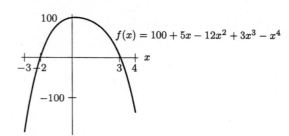

Figure 1.13

  (c) See Figure 1.13. The function has one turning point which is less than the degree of the function.

37. (a) We know that the demand function is of the form

$$q = mp + b$$

where m is the slope and b is the vertical intercept. We know that the slope is

$$m = \frac{q(30) - q(25)}{30 - 25} = \frac{460 - 500}{5} = \frac{-40}{5} = -8.$$

Thus, we get

$$q = -8p + b.$$

Substituting in the point $(30, 460)$ we get

$$460 = -8(30) + b = -240 + b,$$

so that

$$b = 700.$$

Thus, the demand function is

$$q = -8p + 700.$$

(b) We know that the revenue is given by

$$R = pq$$

where p is the price and q is the demand. Thus, we get

$$R = p(-8p + 700) = -8p^2 + 700p.$$

(c) Figure 1.14 shows the revenue as a function of price.

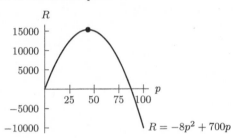

Figure 1.14

Looking at the graph, we see that maximal revenue is attained when the price charged for the product is roughly $44. At this price the revenue is roughly $15,300.

41. The graphs in Figure 1.15 show that in this window, the functions do not look similar.

Comparing these graphs and those in Figure 1.88 shows that the functions look similar on a large scale (far away viewpoint), but not on a smaller scale (close up viewpoint). The reason is that on a smaller scale, the x^4 term does not dominate the other terms (it is not much larger than the other terms).

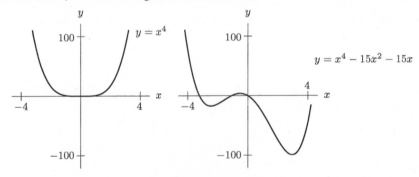

Figure 1.15

Solutions for Section 1.10

1. (a) It makes sense that temperature would be a function of time of day. Assuming similar weather for all the days of the experiment, time of day would probably be the over-riding factor on temperature. Thus, the temperature pattern should repeat itself each day.

(b) The maximum seems to occur at 19 hours or 7 pm. It makes sense that the river would be the hottest toward the end of daylight hours since the sun will have been beating down on it all day. The minimum occurs at around 9 am which also makes sense because the sun has not been up long enough to warm the river much by then.

(c) The period is one day. The amplitude is approximately $\dfrac{32 - 28}{2} = 2°C$

5. Since the volume of the function varies between 2 and 4 liters, the amplitude is 1 and the graph is centered at 3. Since the period is 3, we have

$$3 = \frac{2\pi}{B} \quad \text{so} \quad B = \frac{2\pi}{3}.$$

The correct formula is (b). This function is graphed in Figure 1.16 and we see that it has the right characteristics.

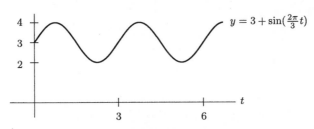

Figure 1.16

9.

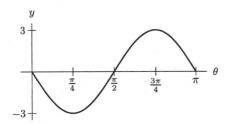

Figure 1.17

See Figure 1.17. The amplitude is 3; the period is π.

13. Let b be the brightness of the star, and t be the time measured in days from a when the star is at its peak brightness. Because of the periodic behavior of the star,

$$b(t) = A\cos(Bt) + C.$$

The oscillations have amplitude $A = 0.35$, shift $C = 4.0$, and period 5.4, so $5.4B = 2\pi$ and $B = 2\pi/5.4$. This gives

$$b(t) = 0.35\cos\left(\frac{2\pi}{5.4}t\right) + 4.$$

17. **(a)** The period looks to be about 12 months. This means that the number of mumps cases oscillates and repeats itself approximately once a year.

$$\text{Amplitude} = \frac{\text{max} - \text{min}}{2} = \frac{11,000 - 2000}{2} = 4500 \text{ cases}$$

This means that the minimum and maximum number of cases of mumps are within $9000 (= 4500 \cdot 2)$ cases of each other.

(b) Assuming cyclical behavior, the number of cases in 30 months will be the same as the number of cases in 6 months which is about 2000 cases. (30 months equals 2 years and six months). The number of cases in 45 months equals the number of cases in 3 years and 9 months which assuming cyclical behavior is the same as the number of cases in 9 months. This is about 2000 as well.

21. This graph is a cosine curve with period 6π and amplitude 5, so it is given by $f(x) = 5\cos\left(\dfrac{x}{3}\right)$.

25. This graph is an inverted cosine curve with amplitude 8 and period 20π, so it is given by $f(x) = -8\cos\left(\dfrac{x}{10}\right)$.

29. This graph has period 8, amplitude 3, and a vertical shift of 3 with no horizontal shift. It is given by

$$f(x) = 3 + 3\sin\left(\frac{2\pi}{8}x\right) = 3 + 3\sin\left(\frac{\pi}{4}x\right).$$

33. (a)

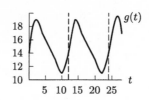

Figure 1.18

From the graph in Figure 1.18, the period appears to be about 12, and the table tells us that $g(0) = 14 = g(12) = g(24)$, which supports this guess.

$$\text{Amplitude} \approx \frac{\text{max} - \text{min}}{2} \approx \frac{19 - 11}{2} = 4$$

(b) To estimate $g(34)$, notice that $g(34) = g(24 + 10) = g(10) = 11$ (assuming the function is periodic with period 12). Likewise, $g(60) = g(24 + 12 + 12 + 12) = 14$.

Solutions for Chapter 1 Review

1. (a) The story in (a) matches Graph (IV), in which the person forgot her books and had to return home.
   (b) The story in (b) matches Graph (II), the flat tire story. Note the long period of time during which the distance from home did not change (the horizontal part).
   (c) The story in (c) matches Graph (III), in which the person started calmly but sped up later.
        The first graph (I) does not match any of the given stories. In this picture, the person keeps going away from home, but his speed decreases as time passes. So a story for this might be: *I started walking to school at a good pace, but since I stayed up all night studying calculus, I got more and more tired the farther I walked.*

5. (a) At $p = 0$, we see $r = 8$. At $p = 3$, we see $r = 7$.
   (b) When $p = 2$, we see $r = 10$. Thus, $f(2) = 10$.

9. The equation of the line is of the form $y = b + mx$ where m is the slope given by

$$\text{Slope} = \frac{\text{Rise}}{\text{Run}} = \frac{2 - 3}{2 - (-1)} = -\frac{1}{3}$$

so the line is of the form $y = b - x/3$. For the line to pass through $(-1, 3)$ we need $3 = b + 1/3$ so $b = 8/3$. Therefore, the equation of the line passing through $(-1, 3)$ and $(2, 2)$ is

$$y = \frac{8}{3} - \frac{1}{3}x.$$

Note that when $x = -1$ this gives $y = 3$ and when $x = 2$ it gives $y = 2$ as required.

13. We know that the linear approximation to this function must take the form

$$P = b + mt$$

where t is in years since 2000. The slope for the function is

$$\text{Slope} = \frac{\Delta P}{\Delta t} = 0.4.$$

We also know that the function takes the value 11.3 in the year 2000 (i.e., at $t = 0$). Thus, if our function is to give the percentage in the years after 2000, it has the form

$$P = 11.3 + 0.4t.$$

17. See Figure 1.19.

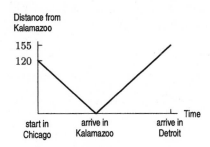

Distance from
Kalamazoo

155

120

Time

start in
Chicago

arrive in
Kalamazoo

arrive in
Detroit

Figure 1.19

21. (a) The average rate of change is the change in attendance divided by the change in time. Between 1999 and 2003,

$$\text{Average rate of change} = \frac{21.64 - 20.76}{2003 - 1999} = 0.22$$

$$= 220{,}000 \text{ people per year.}$$

(b) For each of the years from 1999–2003, the annual increase in the number of games was:

$$1999 \text{ to } 2000 : 20.95 - 20.76 = 0.19$$

$$2000 \text{ to } 2001 : -0.36$$

$$2001 \text{ to } 2002 : 0.92$$

$$2002 \text{ to } 2003 : 0.13.$$

(c)

$$\text{Average of the four figures in part (b)} = \frac{0.19 + (-0.36) + 0.92 + 0.13}{4}$$

$$= \frac{0.88}{4} = 0.22, \text{ which is the same as part (a).}$$

25. This is a line with slope $-3/7$ and y-intercept 3, so a possible formula is

$$y = -\frac{3}{7}x + 3.$$

29. $z = 1 - \cos\theta$

33. This looks like the graph of $y = -x^2$ shifted up 2 units and to the left 3 units. One possible formula is $y = -(x+3)^2 + 2$. Other answers are possible. You can check your answer by graphing it using a calculator or computer.

37. We use logarithms:

$$\ln(e^{5x}) = \ln(100)$$

$$5x = \ln(100)$$

$$x = \frac{\ln(100)}{5} = 0.921.$$

41. (a) We know that $f(x) = 2x + 3$ and $g(x) = \ln x$. Thus,

$$g(f(x)) = g(2x + 3) = \ln(2x + 3).$$

(b) We know that $f(x) = 2x + 3$ and $g(x) = \ln x$. Thus,

$$f(g(x)) = f(\ln x) = 2\ln x + 3.$$

(c) We know that $f(x) = 2x + 3$. Thus,

$$f(f(x)) = f(2x + 3) = 2(2x + 3) + 3 = 4x + 6 + 3 = 4x + 9.$$

45. $m(z) - m(z - h) = z^2 - (z - h)^2 = 2zh - h^2$.

49. (a) We know that regardless of the number of rides one takes, one must pay \$7 to get in. After that, for each ride you must pay another \$1.50, thus the function $R(n)$ is

$$R(n) = 7 + 1.5n.$$

(b) Substituting in the values $n = 2$ and $n = 8$ into our formula for $R(n)$ we get

$$R(2) = 7 + 1.5(2) = 7 + 3 = \$10.$$

This means that admission and 2 rides costs \$10.

$$R(8) = 7 + 1.5(8) = 7 + 12 = \$19.$$

This means that admission and 8 rides costs \$19.

53. We know that at the point where the price is \$1 per scoop the quantity must be 240. Thus we can fill in the graph as follows:

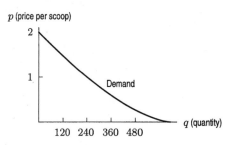

Figure 1.20

(a) Looking at Figure 1.20 we see that when the price per scoop is half a dollar, the quantity given by the demand curve is roughly 360 scoops.

(b) Looking at Figure 1.20 we see that when the price per scoop is \$1.50, the quantity given by the demand curve is roughly 120 scoops.

57. We know that if the population increases exponentially the formula for the population at time t is of the form

$$P(t) = P_0 e^{rt}$$

where P_0 is the population at time $t = 0$ and r is the rate of growth; we can measure time in years. If we let 1980 be the initial time $t = 0$ we get

$$P_0 = 4.478$$

so

$$P(t) = 4.478 e^{rt}.$$

We also know that in 1991, time $t = 11$, we have

$$P(11) = 5.423.$$

Thus we get

$$5.423 = P(11)$$
$$= 4.478 e^{11r}$$
$$e^{11r} = \frac{5.423}{4.478}$$
$$\ln e^{11r} = \ln\left(\frac{5.423}{4.478}\right)$$
$$r = \frac{\ln(5.423/4.478)}{11} \approx 0.0174.$$

Thus, the formula for the population, assuming exponential growth, is

$$P(t) = 4.478e^{0.0174t} \text{ billion.}$$

Trying our formula for the year 2004, time $t = 24$, we get

$$P(24) = 4.478e^{0.0174(24)}$$
$$= 4.478e^{0.4176}$$
$$\approx 6.80 \text{ billion.}$$

Thus, we see that we were not too far off the mark when we approximated the population by an exponential function.

61. (a) The continuous percent growth rate is 15%.
(b) We want $P_0 a^t = 10e^{0.15t}$, so we have $P_0 = 10$ and $a = e^{0.15} = 1.162$. The corresponding function is

$$P = 10(1.162)^t.$$

(c) Since the base in the answer to part (b) is 1.162, the annual percent growth rate is 16.2%. This annual rate is equivalent to the continuous growth rate of 15%.
(d) When we sketch the graphs of $P = 10e^{0.15t}$ and $P = 10(1.162)^t$ on the same axes, we only see one graph. These two exponential formulas are two ways of representing the same function, so the graphs are the same. See Figure 1.21.

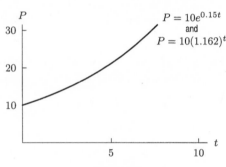

Figure 1.21

65. (a) Since $P(t)$ is an exponential function, it will be of the form $P(t) = P_0 a^t$. We have $P_0 = 1$, since 100% is present at time $t = 0$, and $a = 0.975$, because each year 97.5% of the contaminant remains. Thus,

$$P(t) = (0.975)^t.$$

(b) The graph is shown in Figure 1.22.

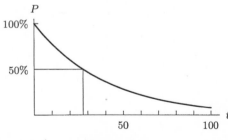

Figure 1.22

(c) The half-life is about 27 years, since $(0.975)^{27} \approx 0.5$.
(d) At time $t = 100$ there is about 8% remaining, since $(0.975)^{100} \approx 0.08$.

69. (a) Option 1 is the best option since money received now can earn interest.
   (b) In Option 1, the entire $2000 earns 5% interest for 1 year, so we have:

$$\text{Future value of Option 1} = 2000e^{0.05 \cdot 1} = \$2102.54.$$

In Option 2, the payment in one year does not earn interest, but the payment made now earns 5% interest for one year. We have

$$\text{Future value of Option 2} = 1000 + 1000e^{0.05 \cdot 1} = 1000 + 1051.27 = \$2051.27.$$

Since Option 3 is paid all in the future, we have

$$\text{Future value of Option 3} = \$2000.$$

As we expected, Option 1 has the highest future value.
   (c) In Option 1, the entire $2000 is paid now, so we have:

$$\text{Present value of Option 1} = \$2000.$$

In Option 2, the payment now has a present value of $1000 but the payment in one year has a lower present value. We have
$$\text{Present value of Option 2} = 1000 + 1000e^{-0.05 \cdot 1} = 1000 + 951.23 = \$1951.23.$$

Since Option 3 is paid all in the future, we have

$$\text{Present value of Option 3} = 2000e^{-0.05 \cdot 1} = \$1902.46.$$

Again, we see that Option 1 has the highest value.
Alternatively, we could have computed the present values directly from the future values found in part (b).

73. 200 revolutions per minute is $\frac{1}{200}$ minutes per revolution, so the period is $\frac{1}{200}$ minutes, or 0.3 seconds.

77. The US voltage has a maximum value of 156 volts and has a period of $1/60$ of a second, so it executes 60 cycles a second.
The European voltage has a higher maximum of 339 volts, and a slightly longer period of $1/50$ seconds, so it oscillates at 50 cycles per second.

Solutions to Problems on Fitting Formulas to Data

1. (a) See Figure 1.23. The regression line fits the data reasonably well.
   (b) The slope is 0.734 trillion dollars per year. The model says that gross world product has been increasing at a rate of approximately 0.734 trillion dollars per year.
   (c) The year 2005 corresponds to $t = 55$ and we have $G = 3.543 + 0.734(55) = 43.913$. Predicted gross world product in 2005 is 43.913 trillion dollars. The year 2020 corresponds to $t = 70$ and we have $G = 3.543 + 0.734(70) = 54.923$. Predicted gross world product in 2020 is 54.923 trillion dollars. We should have much more confidence in the 2005 prediction because it is closer to the time of the data values used.

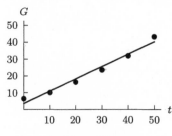

Figure 1.23

5. (a) Using a calculator, we get the stride rate S as a function of speed v,

$$S = 0.08v + 1.77.$$

(b) See Figure 1.24. The line seems to fit the data well.

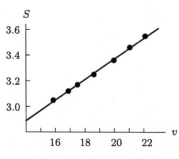

Figure 1.24

(c) Using the formula, we substitute in the given speed of 18 ft/sec into our regression line and get

$$S = (0.08)(18) + 1.77 = 3.21.$$

To find the stride rate when the speed is 10 ft/sec, we substitute $v = 10$ and get

$$S = (0.08)(10) + 1.77 = 2.57.$$

We can have more confidence in our estimate for the stride at the speed 18 ft/sec rather than our estimate at 10 ft/sec, because the speed 18 ft/sec lies well within the domain of the data set given, and thus involves interpolation, whereas the speed 10 ft/sec lies well to the left of the plotted data points and is thus a more speculative value, involving extrapolation.

9. (a) Let P be the population in millions. One algorithm used by a calculator or computer gives

$$P = 180.5(1.011)^t.$$

See Figures 1.25 and 1.26.

(b) The population was growing, on average, at 1.1% per year.

(c) In 2020, we have $t = 60$. The fitted exponential function predicts the population in 2020 to be

$$P = 180.5(1.011)^{60} = 348.0 \text{ million}.$$

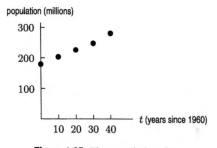

Figure 1.25: The population data

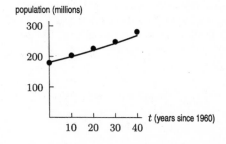

Figure 1.26: The curve $P = 180.5(1.011)^t$ and the data

13. (a) The data is plotted in Figure 1.27.

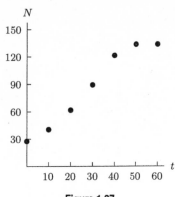

Figure 1.27   **Figure 1.28**

(b) It is not obvious which model is best.

(c) The regression line is

$$N = 2.02t + 26.26.$$

The line and the data are graphed in Figure 1.28. The estimated number of cars in the year 2010, according to this model is, substituting $t = 70$,

$$N = 2.02 \cdot 70 + 26.26 = 167.66 \text{ million.}$$

(d) The slope represents the fact that every year the number of passenger cars in the US grows by approximately 2,020,000.

(e) An exponential regression function that fits the data is

$$N = 32.4(1.028)^t.$$

(Different algorithms can give different formulas, so answers may vary.) The exponential curve and the data are graphed in Figure 1.29.

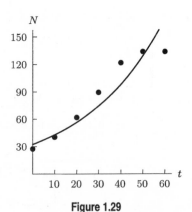

Figure 1.29

The estimated number of cars in the year 2010, according to this model is

$$N = 32.4(1.028)^{70} = 223.9 \text{ million.}$$

This is much larger than the prediction obtained from the linear model.

(f) The annual growth rate is as $a - 1$ for the exponential function $N = N_0 a^t$. Thus, the rate is 0.028 so the annual rate of growth in the number of US passenger cars is 2.8% according to our exponential model.

17. There are several possible answers to this questions depending on whether or not you choose to "smooth out" the given curve. In all cases the leading coefficient is positive. If you did not "smooth out" the function, the graph changes direction 10 times so the degree of the polynomial would be at least 11. If you "smoothed" the function so that it changes direction 4 times, the function would have at least degree 5. Finally, if you "smoothed out" the function so that it only changes direction twice, the degree would be at least 3.

21. **(a)** See Figure 1.30. The data is increasing and then decreasing. The best choice appears to be quadratic regression with negative leading coefficient.

(b) Using a calculator or computer, we have

$$N = -0.0886t^2 + 3.93t + 17.7.$$

(Since different algorithms can give different formulas, answers may vary.)

(c) The year 2005 corresponds to $t = 45$ and we have $N = -0.0886(45)^2 + 3.93(45) + 17.7 = 15.135$. The model predicts that the number of nuclear warheads in the world in the year 2005 will be 15,135.

(d) See Figure 1.31. The quadratic regression equation fits the data reasonably well.

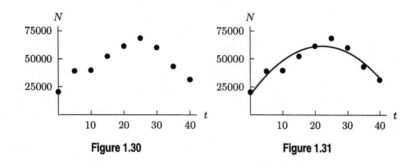

Figure 1.30   Figure 1.31

Solutions to Problems on Compound Interest and the Number e

1. The interest charged on the first month's debt incurs additional interest in the second and subsequent months. The total interest is greater than $12 \times 2\% = 24\%$.

    At the end of 1 year, a debt of \$1 becomes a debt of $(1.02)^{12} \approx \$1.27$, so the annual percentage rate is approximately 27%.

5. In one year, an investment of P_0 becomes $P_0 e^{0.06}$. Using a calculator, we see that

$$P_0 e^{0.06} = P_0(1.0618365)$$

So the effective annual yield is about 6.18%.

9. **(a)** Substitute $n = 10,000$, then $n = 100,000$ and $n = 1,000,000$.

$$\left(1 + \frac{0.04}{10,000}\right)^{10,000} \approx 1.0408107$$

$$\left(1 + \frac{0.04}{100,000}\right)^{100,000} \approx 1.0408108$$

$$\left(1 + \frac{0.04}{1,000,000}\right)^{1,000,000} \approx 1.0408108$$

Effective annual yield:
4.08108%

(b) Evaluating gives $e^{0.04} \approx 1.048108$ as expected.

13. **(a)** Its cost in 1989 would be $1000 + \frac{1290}{100} \cdot 1000 = 13,900$ cruzados.

(b) The monthly inflation rate r solves

$$\left(1 + \frac{1290}{100}\right)^1 = \left(1 + \frac{r}{100}\right)^{12},$$

since we compound the monthly inflation 12 times to get the yearly inflation. Solving for r, we get $r = 24.52\%$. Notice that this is much different than $\frac{1290}{12} = 107.5\%$.

Solutions to Problems on Limits to Infinity and End Behavior ────────

1. As $x \to \infty$, a power function with a larger exponent will dominate a power function with a smaller exponent. The function with the largest values is $y = 0.1x^4$ and the function with the smallest values is $y = 1000x^2$. A global picture is shown in Figure 1.32.

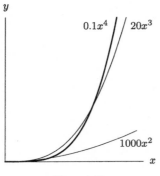

Figure 1.32

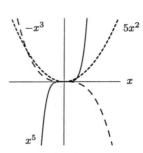

Figure 1.33

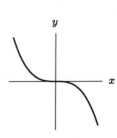

Figure 1.34

5. As $x \to \infty$, $f(x) = x^5$ has the largest positive values. As $x \to -\infty$, $g(x) = -x^3$ has the largest positive values. See Figure 1.33.

9. A possible graph is shown in Figure 1.34. There are many possible answers.

13. The values in Table 1.4 suggest that this limit is 0. The graph of $y = 1/x$ in Figure 1.35 suggests that $y \to 0$ as $x \to \infty$ and so supports the conclusion.

Table 1.4

x	100	1000	1,000,000
$1/x$	0.01	0.001	0.000001

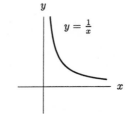

Figure 1.35

17. We see from a graph of $f(x) = 8(1 - e^{-x})$ that $\lim_{x \to \infty} f(x) = 8$ and $\lim_{x \to -\infty} f(x) = -\infty$.

21. A power function with a positive exponent always grows faster than $\ln x$, so the answer is $x^{1/2}$.

25. As $x \to \infty$, a power function with a larger exponent will dominate a power function with a smaller exponent. We see that A corresponds to $y = 0.2x^5$, B corresponds to $y = x^4$, C corresponds to $y = 5x^3$, and D corresponds to $y = 70x^2$.

29.

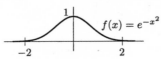

Figure 1.36

   (a) The function f is increasing for negative x. The function f is decreasing for all positive x.
   (b) The graph of f is concave down near $x = 0$.
   (c) As $x \to \infty$, $f(x) \to 0$. As $x \to -\infty$, $f(x) \to 0$.

CHAPTER TWO

Solutions for Section 2.1

1. (a) The average velocity between $t = 3$ and $t = 5$ is

$$\frac{\text{Distance}}{\text{Time}} = \frac{s(5) - s(3)}{5 - 3} = \frac{25 - 9}{2} = \frac{16}{2} = 8 \text{ ft/sec.}$$

(b) Using an interval of size 0.1, we have

$$\left(\begin{array}{c} \text{Instantaneous velocity} \\ \text{at } t = 3 \end{array} \right) \approx \frac{s(3.1) - s(3)}{3.1 - 3} = \frac{9.61 - 9}{0.1} = 6.1.$$

Using an interval of size 0.01, we have

$$\left(\begin{array}{c} \text{Instantaneous velocity} \\ \text{at } t = 3 \end{array} \right) \approx \frac{s(3.01) - s(3)}{3.01 - 3} = \frac{9.0601 - 9}{0.01} = 6.01.$$

From this we guess that the instantaneous velocity at $t = 3$ is about 6 ft/sec.

5. Since 2006 is 6 years after 2000, the rate of growth in 2006 is the derivative of $P(t)$ at $t = 6$. To estimate $P'(t)$ at $t = 6$, we take the interval between $t = 6$ and $t = 6.001$.

$$P'(6) \approx \frac{P(6.001) - P(6)}{6.001 - 6} = \frac{570(1.037)^{6.001} - 570(1.037)^6}{0.001}$$
$$= 25.754 \text{ thousand people/year.}$$

$$P'(6) \approx 25{,}754 \text{ people per year.}$$

9. (a) $f'(x)$ is negative when the function is decreasing and positive when the function is increasing. Therefore, $f'(x)$ is positive at C and G. $f'(x)$ is negative at A and E. $f'(x)$ is zero at B, D, and F.

(b) $f'(x)$ is the largest when the graph of the function is increasing the fastest (i.e. the point with the steepest positive slope). This occurs at point G. $f'(x)$ is the most negative when the graph of the function is decreasing the fastest (i.e. the point with the steepest negative slope). This occurs at point A.

13. (a) From Figure 2.1 we can see that for $x = 1$ the value of the function is decreasing. Therefore, the derivative of $f(x)$ at $x = 1$ is negative.

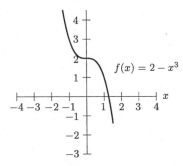

Figure 2.1

(b) $f'(1)$ is the derivative of the function at $x = 1$. This is the rate of change of $f(x) = 2 - x^3$ at $x = 1$. We estimate this by computing the average rate of change of $f(x)$ over intervals near $x = 1$.

Using the intervals $0.999 \leq x \leq 1$ and $1 \leq x \leq 1.001$, we see that

$$\left(\begin{array}{c} \text{Average rate of change} \\ \text{on } 0.999 \leq x \leq 1 \end{array} \right) = \frac{[2 - 1^3] - [2 - 0.999^3]}{1 - 0.999} = \frac{1 - 1.002997}{0.001} = -2.997,$$

$$\left(\begin{array}{c} \text{Average rate of change} \\ \text{on } 1 \leq x \leq 1.001 \end{array} \right) = \frac{[2 - 1.001^3] - [2 - 1^3]}{1.001 - 1} = \frac{0.996997 - 1}{0.001} = -3.003.$$

It appears that the rate of change of $f(x)$ at $x = 1$ is approximately -3, so we estimate $f'(1) = -3$.

17. (a) Since the values of P go up as t goes from 4 to 6 to 8, we see that $f'(6)$ appears to be positive. The percent of households with cable television is increasing at $t = 6$.

(b) We estimate $f'(2)$ using the difference quotient for the interval to the right of $t = 2$, as follows:

$$f'(2) \approx \frac{\Delta P}{\Delta t} = \frac{63.4 - 61.5}{4 - 2} = \frac{1.9}{2} = 0.95.$$

The fact that $f'(2) = 0.95$ tells us that the percent of households with cable television in the United States was increasing at a rate of 0.95 percentage points per year when $t = 2$ (that means 1992).
Similarly:

$$f'(10) \approx \frac{\Delta P}{\Delta t} = \frac{68.9 - 67.8}{12 - 10} = \frac{1.1}{2} = 0.55.$$

The fact that $f'(10) = 0.55$ tells us that the percent of households in the United States with cable television was increasing at a rate of 0.55 percentage points per year when $t = 10$ (that means 2000).

21. (a) Since the point $B = (2, 5)$ is on the graph of g, we have $g(2) = 5$.

(b) The slope of the tangent line touching the graph at $x = 2$ is given by

$$\text{Slope} = \frac{\text{Rise}}{\text{Run}} = \frac{5 - 5.02}{2 - 1.95} = \frac{-0.02}{0.05} = -0.4.$$

Thus, $g'(2) = -0.4$.

25.

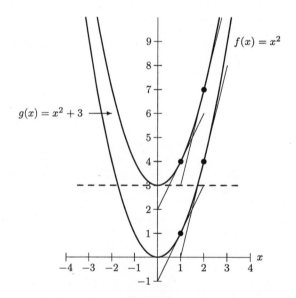

Figure 2.2

(a) The tangent line to the graph of $f(x) = x^2$ at $x = 0$ coincides with the x-axis and therefore is horizontal (slope $= 0$). The tangent line to the graph of $g(x) = x^2 + 3$ at $x = 0$ is the dashed line indicated in the figure and it also has a slope equal to zero. Therefore both tangent lines at $x = 0$ are parallel.

We see in Figure 2.2 that the tangent lines at $x = 1$ appear parallel, and the tangent lines at $x = 2$ appear parallel. The slopes of the tangent lines at any value $x = a$ will be equal.

(b) Adding a constant shifts the graph vertically, but does not change the slope of the curve.

Solutions for Section 2.2

1. Estimating the slope of the lines in Figure 2.3, we find that $f'(-2) \approx 1.0$, $f'(-1) \approx 0.3$, $f'(0) \approx -0.5$, and $f'(2) \approx -1$.

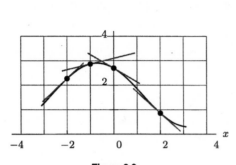

Figure 2.3

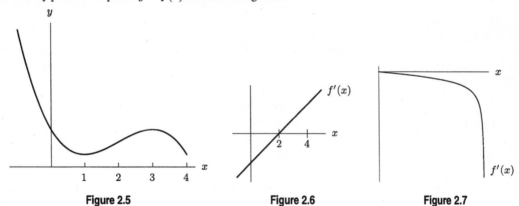

Figure 2.4

5. The slope of this curve is approximately -1 at $x = -4$ and at $x = 4$, approximately 0 at $x = -2.5$ and $x = 1.5$, and approximately 1 at $x = 0$. See Figure 2.4.

9. The function is decreasing for $x < -2$ and $x > 2$, and increasing for $-2 < x < 2$. The matching derivative must be negative (below the x-axis) for $x < -2$ and $x > 2$, positive (above the x-axis) for $-2 < x < 2$, and zero (on the x-axis) for $x = -2$ and $x = 2$. The matching derivative is in graph VIII.

13. (a) x_3 (b) x_4 (c) x_5 (d) x_3

17. Since $f'(x) > 0$ for $1 < x < 3$, $f(x)$ is increasing on this interval.
Since $f'(x) < 0$ for $x < 1$ or $x > 3$, $f(x)$ is decreasing on these intervals.
Since $f'(x) = 0$ for $x = 1$ and $x = 3$, the tangent to $f(x)$ will be horizontal at these x's.
One of many possible shapes of $y = f(x)$ is shown in Figure 2.5.

Figure 2.5   **Figure 2.6**   **Figure 2.7**

21. See Figure 2.6.

25. See Figure 2.7.

29. (a) The function f is increasing where f' is positive, so for $x_1 < x < x_3$.
(b) The function f is decreasing where f' is negative, so for $0 < x < x_1$ or $x_3 < x < x_5$.

Solutions for Section 2.3

1. kilograms/meter

5. (a) The statement $f(5) = 18$ means that when 5 milliliters of catalyst are present, the reaction will take 18 minutes. Thus, the units for 5 are ml while the units for 18 are minutes.
(b) As in part (a), 5 is measured in ml. Since f' tells how fast T changes per unit a, we have f' measured in minutes/ml. If the amount of catalyst increases by 1 ml (from 5 to 6 ml), the reaction time decreases by about 3 minutes.

9. (a) Positive, since weight increases as the child gets older.
(b) $f(8) = 45$ tells us that when the child is 8 years old, the child weighs 45 pounds.
(c) The units of $f'(a)$ are lbs/year. $f'(a)$ tells the rate of growth in lbs/years at age a.
(d) $f'(8) = 4$ tells us that the 8-year-old child is growing at about 4 lbs/year.
(e) As a increases, $f'(a)$ will decrease since the rate of growth slows down as the child grows up.

13. Compare the secant line to the graph from week 0 to week 40 to the tangent lines at week 20 and week 36.

  (a) At week 20 the secant line is steeper than the tangent line. The instantaneous weight growth rate is less than the average.

  (b) At week 36 the tangent line is steeper than the secant line. The instantaneous weight growth rate is greater than the average.

17. Since $f(t) = 1.291(1.006)^t$, we have

$$f(6) = 1.291(1.006)^6 = 1.338.$$

To estimate $f'(6)$, we use a small interval around 6:

$$f'(6) \approx \frac{f(6.001) - f(6)}{6.001 - 6} = \frac{1.291(1.006)^{6.001} - 1.291(1.006)^6}{0.001} = 0.008.$$

We see that $f(6) = 1.338$ billion people and $f'(6) = 0.008$ billion (that is, 8 million) people per year. This model predicts that the population of China will be about 1,338,000,000 people in 2009 and growing at a rate of about 8,000,000 people per year at that time.

21. Using $\Delta x = 1$, we can say that

$$\begin{aligned} f(21) &= f(20) + \text{change in } f(x) \\ &\approx f(20) + f'(20)\Delta x \\ &= 68 + (-3)(1) \\ &= 65. \end{aligned}$$

Similarly, using $\Delta x = -1$,

$$\begin{aligned} f(19) &= f(20) + \text{change in } f(x) \\ &\approx f(20) + f'(20)\Delta x \\ &= 68 + (-3)(-1) \\ &= 71. \end{aligned}$$

Using $\Delta x = 5$, we can write

$$\begin{aligned} f(25) &= f(20) + \text{change in } f(x) \\ &\approx f(20) + (-3)(5) \\ &= 68 - 15 \\ &= 53. \end{aligned}$$

25. (a) The tangent line is shown in Figure 2.8. Two points on the line are $(0, 16)$ and $(3.2, 0)$. Thus

$$\text{Slope} = \frac{0 - 16}{3.2 - 0} = -5 \text{ (cm/sec)/kg}.$$

  (b) Since 50 grams = 0.050 kg, the contraction velocity changes by about $-5(\text{cm/sec})/\text{kg} \cdot 0.050\text{kg} = -0.25$ cm/sec. The velocity is reduced by 0.25 cm/sec or 2.5 mm/sec.

  (c) Since $v(x)$ is the contraction velocity in cm/sec with a load of x kg, we have $v'(2) = -5$.

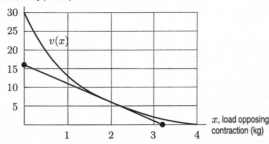

Figure 2.8

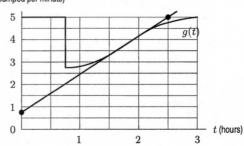

Figure 2.9

29. (a) The tangent line is shown in Figure 2.9. Two points on the line are $(0, 0.75)$ and $(2.5, 5)$. The

$$\text{Slope} = \frac{5 - 0.75}{2.5 - 0} = 1.7 \text{ (liters/minute)/hour.}$$

(b) The rate of change of the pumping rate is the slope of the tangent line. One minute = $1/60$ hour, so in one minute the

$$\text{Pumping rate increases by about } 1.7 \frac{\text{(liter/minute)}}{\text{hour}} \cdot \frac{1}{60} \text{ hour} = 0.028 \text{ liter/minute.}$$

(c) Since $g(t)$ is the pumping rate in liters/minute at time t hours, we have $g'(2) = 1.7$.

33. The fat consumption rate (kg/week) is the rate at which the quantity of fat is decreasing, that is, -1 times the derivative of the fat storage function. We estimate the derivatives at 3, 6, and 8 weeks by drawing tangent lines to the storage graph, shown in Figure 2.10, and calculating their slopes.

(a) The tangent line at 3 weeks is the storage graph itself since that part of the graph is straight. Two points on the tangent line are $(0, 12)$ and $(4, 4)$. Then

$$\text{Slope} = \frac{4 - 12}{4 - 0} = -2.0 \text{ kg/week.}$$

The consumption rate is 2.0 kg/week.

(b) Two points on the tangent line are $(0, 4.7)$ and $(8, 0.1)$. Then,

$$\text{Slope} = \frac{0.1 - 4.7}{8 - 0} = -0.6 \text{ kg/week.}$$

The consumption rate is 0.6 kg/week.

(c) Two points on the tangent line are $(0, 2.7)$ and $(8, 0.6)$. Then

$$\text{Slope} = \frac{0.6 - 2.7}{8 - 0} = -0.3 \text{ kg/week.}$$

The consumption rate is 0.3 kg/week.

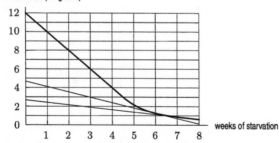

quantities of stored fat (kilograms)

Figure 2.10: Rates of consumption are -1 times slopes of tangent lines

37. (a) The units of compliance are units of volume per units of pressure, or liters per centimeter of water.
(b) The increase in volume for a 5 cm reduction in pressure is largest between 10 and 15 cm. Thus, the compliance appears maximum between 10 and 15 cm of pressure reduction. The derivative is given by the slope, so

$$\text{Compliance} \approx \frac{0.70 - 0.49}{15 - 10} = 0.042 \text{ liters per centimeter.}$$

(c) When the lung is nearly full, it cannot expand much more to accommodate more air.

Solutions for Section 2.4

1. (a) Since the graph is below the x-axis at $x = 2$, the value of $f(2)$ is negative.
(b) Since $f(x)$ is decreasing at $x = 2$, the value of $f'(2)$ is negative.
(c) Since $f(x)$ is concave up at $x = 2$, the value of $f''(2)$ is positive.

5. $f'(x) < 0$
$f''(x) > 0$

9. The derivative, $s'(t)$, appears to be positive since $s(t)$ is increasing over the interval given. The second derivative also appears to be positive or zero since the function is concave up or possibly linear between $t = 1$ and $t = 3$, i.e., it is increasing at a non-decreasing rate.

13. The two points at which $f' = 0$ are A and B. Since f' is nonzero at C and D and f'' is nonzero at all four points, we get the completed Table 2.1:

Table 2.1

Point	f	f'	f''
A	$-$	0	$+$
B	$+$	0	$-$
C	$+$	$-$	$-$
D	$-$	$+$	$+$

17. (a) The function appears to be decreasing and concave down, and so we conjecture that f' is negative and that f'' is negative.

(b) We use difference quotients to the right:
$f'(2) \approx \frac{137-145}{4-2} = -4$
$f'(8) \approx \frac{56-98}{10-8} = -21.$

21. To the right of $x = 5$, the function starts by increasing, since $f'(5) = 2 > 0$ (though f may subsequently decrease) and is concave down, so its graph looks like the graph shown in Figure 2.11. Also, the tangent line to the curve at $x = 5$ has slope 2 and lies above the curve for $x > 5$. If we follow the tangent line until $x = 7$, we reach a height of 24. Therefore, $f(7)$ must be smaller than 24, meaning 22 is the only possible value for $f(7)$ from among the choices given.

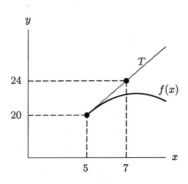

Figure 2.11

25. (a)

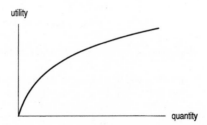

(b) As a function of quantity, utility is increasing but at a decreasing rate; the graph is increasing but concave down. So the derivative of utility is positive, but the second derivative of utility is negative.

29. (a) IV, (b) III, (c) II, (d) I, (e) IV, (f) II

Solutions for Section 2.5

1. Drawing in the tangent line at the point $(10000, C(10000))$ we get Figure 2.12.

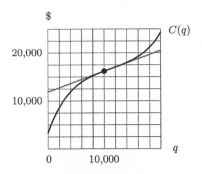

Figure 2.12

We see that each vertical increase of 2500 in the tangent line gives a corresponding horizontal increase of roughly 6000. Thus the marginal cost at the production level of 10,000 units is

$$C'(10,000) = \frac{\text{Slope of tangent line}}{\text{to } C(q) \text{ at } q = 10,000} = \frac{2500}{6000} = 0.42.$$

This tells us that after producing 10,000 units, it will cost roughly $0.42 to produce one more unit.

5. Marginal cost $= C'(q)$. Therefore, marginal cost at q is the slope of the graph of $C(q)$ at q. We can see that the slope at $q = 5$ is greater than the slope at $q = 30$. Therefore, marginal cost is greater at $q = 5$. At $q = 20$, the slope is small, whereas at $q = 40$ the slope is larger. Therefore, marginal cost at $q = 40$ is greater than marginal cost at $q = 20$.

9. (a) The cost to produce 50 units is $4300 and the marginal cost to produce additional items is about $24 per unit. Producing two more units (from 50 to 52) increases cost by $48. We have

$$C(52) \approx 4300 + 24(2) = \$4348.$$

   (b) When $q = 50$, the marginal cost is $24 per item and the marginal revenue is $35 per item. The profit on the 51st item is $35 - 24 = \$11$.

   (c) When $q = 100$, the marginal cost is $38 per item and the marginal revenue is $35 per item, so the company will lose $3 by producing the 101st item. Since the company will lose money, it should not produce the 101st item.

13. (a) At $q = 2.1$ million,

$$\text{Profit} = \pi(2.1) = R(2.1) - C(2.1) = 6.9 - 5.1 = 1.8 \text{ million dollars.}$$

   (b) If $\Delta q = 0.04$,

$$\text{Change in revenue, } \Delta R \approx R'(2.1)\Delta q = 0.7(0.04) = 0.028 \text{ million dollars} = \$28,000.$$

Thus, revenues increase by $28,000.

   (c) If $\Delta q = -0.05$,

$$\text{Change in revenue, } \Delta R \approx R'(2.1)\Delta q = 0.7(-0.05) = -0.035 \text{ million dollars} = -\$35,000.$$

Thus, revenues decrease by $35,000.

   (d) We find the change in cost by a similar calculation. For $\Delta q = 0.04$,

$$\text{Change in cost, } \Delta C \approx C'(2.1)\Delta q = 0.6(0.04) = 0.024 \text{ million dollars} = \$24,000$$
$$\text{Change in profit, } \Delta\pi = \$28,000 - \$24,000 = \$4000.$$

Thus, increasing production 0.04 million units increases profits by $4000.
      For $\Delta q = -0.05$,

$$\text{Change in cost, } \Delta C \approx C'(2.1)\Delta q = 0.6(-0.05) = -0.03 \text{ million dollars} = -\$30,000$$
$$\text{Change in profit, } \Delta\pi = -\$35,000 - (-\$30,000) = -\$5000.$$

Thus, decreasing production 0.05 million units decreases profits by $5000.

Solutions for Chapter 2 Review

1. The slope is positive at A and D; negative at C and F. The slope is most positive at A; most negative at F.

5. Between 1804 and 1927, the world's population increased 1 billion people in 123 years, for an average rate of change of $1/123$ billion people per year. We convert this to people per minute:

$$\frac{1,000,000,000}{123} \text{ people/year} \cdot \frac{1}{60 \cdot 24 \cdot 365} \text{ years/minute} = 15.47 \text{ people/minute.}$$

Between 1804 and 1927, the population of the world increased at an average rate of 15.47 people per minute. Similarly, we find the following:

Between 1927 and 1960, the increase was 57.65 people per minute.
Between 1960 and 1974, the increase was 135.90 people per minute.
Between 1974 and 1987, the increase was 146.35 people per minute.
Between 1987 and 1999, the increase was 158.55 people per minute.

9. This is a line with slope 1, so the derivative is the constant function $f'(x) = 1$. The graph is the horizontal line $y = 1$. See Figure 2.13.

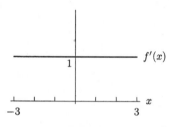

Figure 2.13

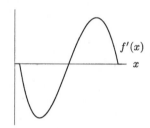

Figure 2.14

13. See Figure 2.14.

17. The statements $f(100) = 35$ and $f'(100) = 3$ tell us that at $x = 100$, the value of the function is 35 and the function is increasing at a rate of 3 units for a unit increase in x. Since we increase x by 2 units in going from 100 to 102, the value of the function goes up by approximately $2 \cdot 3 = 6$ units, so

$$f(102) \approx 35 + 2 \cdot 3 = 35 + 6 = 41.$$

21. (a) This means that investing the $1000 at 5% would yield $1649 after 10 years.
   (b) Writing $g'(r)$ as dB/dr, we see that the units of dB/dr are dollars per percent (interest). We can interpret dB as the extra money earned if interest rate is increased by dr percent. Therefore $g'(5) = \frac{dB}{dr}\big|_{r=5} \approx 165$ means that the balance, at 5% interest, would increase by about $165 if the interest rate were increased by 1%. In other words, $g(6) \approx g(5) + 165 = 1649 + 165 = 1814$.

25. (a) The company hopes that increased advertising always brings in more customers instead of turning them away. Therefore, it hopes $f'(a)$ is always positive.
   (b) If $f'(100) = 2$, it means that if the advertising budget is $100,000, each extra dollar spent on advertising will bring in $2 worth of sales. If $f'(100) = 0.5$, each dollar above $100 thousand spent on advertising will bring in $0.50 worth of sales.
   (c) If $f'(100) = 2$, then as we saw in part (b), spending slightly more than $100,000 will increase revenue by an amount greater than the additional expense, and thus more should be spent on advertising. If $f'(100) = 0.5$, then the increase in revenue is less than the additional expense, hence too much is being spent on advertising. The optimum amount to spend, a, is an amount that makes $f'(a) = 1$. At this point, the increases in advertising expenditures just pay for themselves. If $f'(a) < 1$, too much is being spent; if $f'(a) > 1$, more should be spent.

29. Since all advertising campaigns are assumed to produce an increase in sales, a graph of sales against time would be expected to have a positive slope.

A positive second derivative means the rate at which sales are increasing is increasing. If a positive second derivative is observed during a new campaign, it is reasonable to conclude that this increase in the rate sales are increasing is caused by the new campaign—which is therefore judged a success. A negative second derivative means a decrease in the rate at which sales are increasing, and therefore suggests the new campaign is a failure.

33. (a) The rate of energy consumption required when $v = 0$ is the vertical intercept, about 1.8 joules/sec.
    (b) The graph shows $f(v)$ first decreases and then increases as v increases. This tells us that the bird expends more energy per second to remain still than to travel at slow speeds (say 0.5 to 1 meter/sec), but that the rate of energy consumption required increases again at speeds beyond 1 meter/sec. The upward concavity of the graph tells us that as the bird speeds up, it uses energy at a faster and faster rate.
    (c) Figure 2.15 shows a possible graph of the derivative $f'(v)$. Other answers are possible.

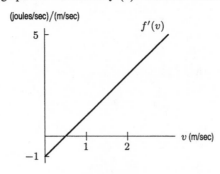

Figure 2.15

Solutions to Problems on Limits and the Definition of the Derivative

1. The answers to parts (a)–(f) are marked in Figure 2.16.

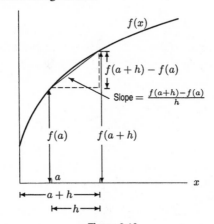

Figure 2.16

5. Using $h = 0.1, 0.01, 0.001$, we see

$$\frac{(3+0.1)^3 - 27}{0.1} = 27.91$$

$$\frac{(3+0.01)^3 - 27}{0.01} = 27.09$$

$$\frac{(3+0.001)^3 - 27}{0.001} = 27.009.$$

These calculations suggest that $\lim_{h \to 0} \dfrac{(3+h)^3 - 27}{h} = 27.$

9. Yes, $f(x)$ is continuous on $0 \le x \le 2$.

13. Yes: $f(x) = x + 2$ is continuous for all values of x.

17. No: $f(x) = 1/(x - 1)$ is not continuous on any interval containing $x = 1$.

21. Since we can't make a fraction of a pair of pants, the number increases in jumps, so the function is not continuous.

25. Using the definition of the derivative, we have

$$
\begin{aligned}
f'(x) &= \lim_{h \to 0} \frac{f(x + h) - f(x)}{h} \\
&= \lim_{h \to 0} \frac{(3(x + h) - 2) - (3x - 2)}{h} \\
&= \lim_{h \to 0} \frac{3x + 3h - 2 - 3x + 2}{h} \\
&= \lim_{h \to 0} \frac{3h}{h}.
\end{aligned}
$$

As h gets very close to zero without actually equaling zero, we can cancel the h in the numerator and denominator to obtain

$$
f'(x) = \lim_{h \to 0} (3) = 3.
$$

29. Using the definition of the derivative, we have

$$
\begin{aligned}
f'(x) &= \lim_{h \to 0} \frac{f(x + h) - f(x)}{h} \\
&= \lim_{h \to 0} \frac{((x + h) - (x + h)^2) - (x - x^2)}{h} \\
&= \lim_{h \to 0} \frac{(x + h - (x^2 + 2xh + h^2)) - x + x^2}{h} \\
&= \lim_{h \to 0} \frac{x + h - x^2 - 2xh - h^2 - x + x^2}{h} \\
&= \lim_{h \to 0} \frac{h - 2xh - h^2}{h}.
\end{aligned}
$$

As long as we let h get close to zero without actually equaling zero, we can cancel the h in the numerator and denominator, and we are left with $1 - 2x - h$. Taking the limit as h goes to zero, we get $f'(x) = 1 - 2x$ since the other term goes to zero.

33. Using the definition of the derivative, we have

$$
\begin{aligned}
f'(x) &= \lim_{h \to 0} \frac{f(x + h) - f(x)}{h} \\
&= \lim_{h \to 0} \frac{1/(x + h) - 1/x}{h}.
\end{aligned}
$$

Writing the numerator over a common denominator and simplifying, we get

$$
\begin{aligned}
f'(x) &= \lim_{h \to 0} \frac{(x - (x + h))/(x(x + h))}{h} \\
&= \lim_{h \to 0} \frac{-h/(x(x + h))}{h} \\
&= \lim_{h \to 0} \frac{-h}{hx(x + h)}.
\end{aligned}
$$

As long as we let h get close to zero without actually equaling zero, we can cancel the h in the numerator and denominator, and we are left with $-1/(x(x + h))$. Taking the limit as h goes to zero, we get $f'(x) = -1/x^2$ since h goes to zero.

CHAPTER THREE

Solutions for Section 3.1

1. $\dfrac{dy}{dx} = 0$

5. $y' = \frac{4}{3}x^{1/3}$.

9. $f'(x) = -4x^{-5}$.

13. $\dfrac{dy}{dx} = 6x + 7$.

17. Since $g(t) = \dfrac{1}{t^5} = t^{-5}$, we have $g'(t) = -5t^{-6}$.

21. Since $h(\theta) = \dfrac{1}{\sqrt[3]{\theta}} = \theta^{-1/3}$, we have $h'(\theta) = -\dfrac{1}{3}\theta^{-4/3}$.

25. $y' = 6t - \frac{6}{t^{3/2}} + \frac{2}{t^3}$.

29. $f'(x) = k \cdot \frac{d}{dx}(x^2) = 2kx$.

33. Since a and b are constants, we have

$$\frac{dP}{dt} = 0 + b\frac{1}{2}t^{-1/2} = \frac{b}{2\sqrt{t}}.$$

37. **(a)** $f'(t) = 2t - 4$.

   (b) $f'(1) = 2(1) - 4 = -2$
   $f'(2) = 2(2) - 4 = 0$

   (c) We see from part (b) that $f'(2) = 0$. This means that the slope of the line tangent to the curve at $x = 2$ is zero. From Figure 3.1, we see that indeed the tangent line is horizontal at the point $(2, 1)$. The fact that $f'(1) = -2$ means that the slope of the line tangent to the curve at $x = 1$ is -2. If we draw a line tangent to the graph at $x = 1$ (the point $(1, 2)$) we see that it does indeed have a slope of -2.

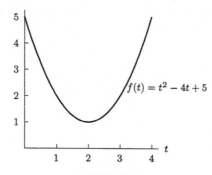

$f(t) = t^2 - 4t + 5$

Figure 3.1

41. The rate of change of the population is given by the derivative. For $P(t) = t^3 + 4t + 1$ the derivative is $P'(t) = 3t^2 + 4$. At $t = 2$, the rate of change of the population is $3(2)^2 + 4 = 12 + 4 = 16$, meaning the population is growing by 16 units per unit of time.

45. **(a)** We have $f(2) = 8$, so a point on the tangent line is $(2, 8)$. Since $f'(x) = 3x^2$, the slope of the tangent is given by

$$m = f'(2) = 3(2)^2 = 12.$$

Thus, the equation is

$$y - 8 = 12(x - 2) \quad \text{or} \quad y = 12x - 16.$$

(b) See Figure 3.2. The tangent line lies below the function $f(x) = x^3$, so estimates made using the tangent line are underestimates.

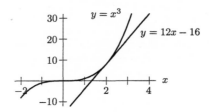

Figure 3.2

49. The marginal cost of producing the 25th item is $C'(25)$, where $C'(q) = 4q$, so the marginal cost is $100. This means that the cost of production increases by about $100 when we add one unit to a production level of 25 units.

53. (a) The marginal cost function describes the change in cost on the margin, i.e. the change in cost associated with each additional item produced. Thus it is the change in cost with respect to change in production (dC/dq). We recognize this as an expression for the derivative of $C(q)$ with respect to q. Thus, the marginal cost function equals $C'(q) = 0.08(3q^2) + 75 = 0.24q^2 + 75$.

(b)
$$C(50) = 0.08(50)^3 + 75(50) + 1000 = \$14,750.$$

$C(50)$ tells us how much it costs to produce 50 items. From above we can see that the company spends $14,750 to produce 50 items, and thus the units for $C(q)$ are dollars.

$$C'(50) = 0.24(50)^2 + 75 = \$675 \text{ per item.}$$

$C'(q)$ tells us the change in cost to produce one additional item of product. Thus at $q = 50$ costs will increase by $675 for each additional item of product produced, and thus the units are dollars/item.

57. If $f(x) = x^n$, then $f'(x) = nx^{n-1}$. This means $f'(1) = n \cdot 1^{n-1} = n \cdot 1 = n$, because any power of 1 equals 1.

Solutions for Section 3.2

1. $f'(x) = 2e^x + 2x$.

5. Since $y = 2^x + \dfrac{2}{x^3} = 2^x + 2x^{-3}$, we have $\dfrac{dy}{dx} = (\ln 2)2^x - 6x^{-4}$.

9. $\dfrac{dy}{dx} = 3 - 2(\ln 4)4^x$.

13. $P'(t) = Ce^t$.

17. $R' = \frac{3}{q}$.

21. $\dfrac{dy}{dx} = 2x + 4 - 3/x$.

25. (a) $f(x) = 1 - e^x$ crosses the x-axis where $0 = 1 - e^x$, which happens when $e^x = 1$, so $x = 0$. Since $f'(x) = -e^x$, $f'(0) = -e^0 = -1$.
(b) $y = -x$

29. Since $P = 1 \cdot (1.05)^t$, $\frac{dP}{dt} = \ln(1.05)1.05^t$. When $t = 10$,

$$\frac{dP}{dt} = (\ln 1.05)(1.05)^{10} \approx \$0.07947/\text{year} \approx 7.95¢/\text{year.}$$

33. (a) For $y = \ln x$, we have $y' = 1/x$, so the slope of the tangent line is $f'(1) = 1/1 = 1$. The equation of the tangent line is $y - 0 = 1(x - 1)$, so, on the tangent line, $y = g(x) = x - 1$.

(b) Using a value on the tangent line to approximate $\ln(1,1)$, we have

$$\ln(1.1) \approx g(1.1) = 1.1 - 1 = 0.1.$$

Similarly, $\ln(2)$ is approximated by

$$\ln(2) \approx g(2) = 2 - 1 = 1.$$

(c) From Figure 3.3, we see that $f(1.1)$ and $f(2)$ are below $g(x) = x - 1$. Similarly, $f(0.9)$ and $f(0.5)$ are also below $g(x)$. This is true for any approximation of this function by a tangent line since f is concave down ($f''(x) = -\frac{1}{x^2} < 0$ for all $x > 0$). Thus, for a given x-value, the y-value given by the function is always below the value given by the tangent line.

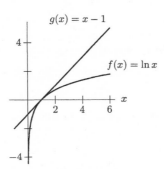

Figure 3.3

Solutions for Section 3.3

1. $\dfrac{d}{dx}\left((4x^2 + 1)^7\right) = 7(4x^2 + 1)^6 \dfrac{d}{dx}(4x^2 + 1) = 7(4x^2 + 1)^6 \cdot 8x = 56x(4x^2 + 1)^6.$

5. $w' = 100(t^3 + 1)^{99}(3t^2) = 300t^2(t^3 + 1)^{99}.$

9. $\dfrac{dy}{dt} = e^{0.7t} \cdot (0.7) = 0.7e^{0.7t}.$

13. $\dfrac{dP}{dt} = 200(0.12)e^{0.12t} = 24e^{0.12t}.$

17. $\dfrac{dy}{dt} = 5(5e^{5t+1}) = 25e^{5t+1}.$

21. $f'(x) = \dfrac{-1}{1 - x} = \dfrac{1}{x - 1}.$

25. $f'(t) = 5 \cdot \dfrac{1}{5t + 1} \cdot 5 = \dfrac{25}{5t + 1}.$

29. $\dfrac{dy}{dx} = 5 + \dfrac{1}{x + 2} \cdot 1 = 5 + \dfrac{1}{x + 2}.$

33. $f'(x) = \dfrac{1}{2}(1 - x^2)^{-\frac{1}{2}}(-2x) = \dfrac{-x}{\sqrt{1 - x^2}}.$

37. To find the equation of the line, we need a point and the slope. Since $f(4) = 10e^{-0.2(4)} = 4.493$, the point on the line is $(4, 4.493)$. We use the derivative to find the slope:

$$f'(x) = 10e^{-0.2x}(-0.2) = -2e^{-0.2x}.$$

Substituting $x = 4$, we see that the slope is

$$m = f'(4) = -2e^{-0.2(4)} = -0.899.$$

Using $y - y_0 = m(x - x_0)$, we find that the equation for the tangent line is:

$$y - 4.493 = -0.899(x - 4)$$

Simplifying, we have

$$y = -0.899x + 8.089.$$

41. $f(p) = 10,000e^{-0.25p}$, $f(2) = 10,000e^{-0.5} \approx 6065$. If the product sells for \$2, then 6065 units can be sold.

$$f'(p) = 10,000e^{-0.25p}(-0.25) = -2500e^{-0.25p}$$

$$f'(2) = -2500e^{-0.5} \approx -1516.$$

Thus, at a price of \$2, a \$1 increase in price results in a decrease in quantity sold of 1516 units .

45. (a) $P(12) = 10e^{0.6(12)} = 10e^{7.2} \approx 13,394$ fish. There are 13,394 fish in the area after 12 months.
   (b) We differentiate to find $P'(t)$, and then substitute in to find $P'(12)$:

$$P'(t) = 10(e^{0.6t})(0.6) = 6e^{0.6t}$$
$$P'(12) = 6e^{0.6(12)} \approx 8037 \text{ fish/month.}$$

The population is growing at a rate of approximately 8037 fish per month.

49. The chain rule gives

$$\frac{d}{dx}f(g(x))\bigg|_{x=30} = f'(g(30))g'(30) = f'(55)g'(30) = (1)(\tfrac{1}{2}) = \frac{1}{2}.$$

53. From the graphs, we estimate $g(1) \approx 2$, $g'(1) \approx 1$, and $f'(2) \approx 0.8$. Thus, by the chain rule,

$$h'(1) = f'(g(1)) \cdot g'(1) \approx f'(2) \cdot g'(1) \approx 0.8 \cdot 1 = 0.8.$$

Solutions for Section 3.4

1. By the product rule,

$$f'(x) = 2(3x - 2) + (2x + 1) \cdot 3 = 12x - 1.$$

Alternatively,

$$f'(x) = (6x^2 - x - 2)' = 12x - 1.$$

The two answers match.

5. Differentiating with respect to x, we have

$$\frac{dy}{dx} = \frac{d}{dx}(5xe^{x^2}) = \left(\frac{d}{dx}(5x)\right)e^{x^2} + 5x\frac{d}{dx}(e^{x^2})$$
$$= (5)e^{x^2} + 5x(e^{x^2} \cdot 2x)$$
$$= 5e^{x^2} + 10x^2e^{x^2}.$$

9. $\dfrac{dz}{dt} = (3t + 1)5 + 3(5t + 2) = 15t + 5 + 15t + 6 = 30t + 11.$
We could have started by multiplying the factors to obtain $15t^2 + 11t + 2$, and then taken the derivative of the result.

13. Divide and then differentiate
$$f(x) = x + \frac{3}{x}$$
$$f'(x) = 1 - \frac{3}{x^2}.$$

17. $g'(p) = p\left(\dfrac{2}{2p+1}\right) + \ln(2p+1)(1) = \dfrac{2p}{2p+1} + \ln(2p+1)$.

21. $w' = (3t^2 + 5)(t^2 - 7t + 2) + (t^3 + 5t)(2t - 7)$.

25. Using the quotient rule gives

$$\frac{dz}{dt} = \frac{d}{dt}\left(\frac{1-t}{1+t}\right) = \frac{-1 \cdot (1+t) - (1-t) \cdot 1}{(1+t)^2} = \frac{-1 - t - 1 + t}{(1+t)^2} = \frac{-2}{(1+t)^2}.$$

29. Since a and b are constants, we have $f'(t) = ae^{bt}(b) = abe^{bt}$.

33. Using the product and chain rules, we have

$$g'(\alpha) = e^{\alpha e^{-2\alpha}} \cdot \frac{d}{dx}(\alpha e^{-2\alpha}) = e^{\alpha e^{-2\alpha}}\left(1 \cdot e^{-2\alpha} + \alpha e^{-2\alpha}(-2)\right)$$
$$= e^{\alpha e^{-2\alpha}}(e^{-2\alpha} - 2\alpha e^{-2\alpha})$$
$$= (1 - 2\alpha)e^{-2\alpha}e^{\alpha e^{-2\alpha}}.$$

37. **(a)** $q(10) = 5000e^{-0.8} \approx 2247$ units.

(b) $q' = 5000(-0.08)e^{-0.08p} = -400e^{-0.08p}$, $q'(10) = -400e^{-0.8} \approx -180$. This means that at a price of $10, a $1 increase in price will result in a decrease in quantity demanded by 180 units.

41. **(a)** See Figure 3.4.

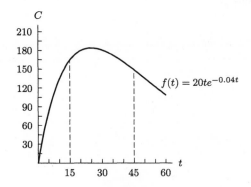

Figure 3.4

Looking at the graph of C, we can see that the see that at $t = 15$, C is increasing. Thus, the slope of the curve at that point is positive, and so $f'(15)$ is also positive. At $t = 45$, the function is decreasing, i.e. the slope of the curve is negative, and thus $f'(45)$ is negative.

(b) We begin by differentiating the function:

$$f'(t) = (20t)(-0.04e^{-0.04t}) + (e^{-0.04t})(20)$$
$$f'(t) = e^{-0.04t}(20 - 0.8t).$$

At $t = 30$,

$$f(30) = 20(30)e^{-0.04\cdot(30)} = 600e^{-1.2} \approx 181 \text{ mg/ml}$$
$$f'(30) = e^{-1.2}(20 - (0.8)(30)) = e^{-1.2}(-4) \approx -1.2 \text{ mg/ml/min}.$$

These results mean the following: At $t = 30$, or after 30 minutes, the concentration of the drug in the body ($f(30)$) is about 181 mg/ml. The rate of change of the concentration ($f'(30)$) is about -1.2 mg/ml/min, meaning that the concentration of the drug in the body is dropping by 1.2 mg/ml each minute at $t = 30$ minutes.

45. (a) We multiply through by $h = f \cdot g$ and cancel as follows:

$$\frac{f'}{f} + \frac{g'}{g} = \frac{h'}{h}$$

$$\left(\frac{f'}{f} + \frac{g'}{g}\right) \cdot fg = \frac{h'}{h} \cdot fg$$

$$\frac{f'}{f} \cdot fg + \frac{g'}{g} \cdot fg = \frac{h'}{h} \cdot h$$

$$f' \cdot g + g' \cdot f = h',$$

which is the product rule.

(b) We start with the product rule, multiply through by $1/(fg)$ and cancel as follows:

$$f' \cdot g + g' \cdot f = h'$$

$$(f' \cdot g + g' \cdot f) \cdot \frac{1}{fg} = h' \cdot \frac{1}{fg}$$

$$(f' \cdot g) \cdot \frac{1}{fg} + (g' \cdot f) \cdot \frac{1}{fg} = h' \cdot \frac{1}{fg}$$

$$\frac{f'}{f} + \frac{g'}{g} = \frac{h'}{h},$$

which is the additive rule shown in part (a).

Solutions for Section 3.5

1. $\dfrac{dy}{dx} = 5 \cos x.$

5. $R'(q) = 2q + 2\sin q.$

9. $\dfrac{dW}{dt} = 4(-\sin(t^2)) \cdot 2t = -8t\sin(t^2).$

13. $z' = -4\sin(4\theta).$

17. $f'(\theta) = 3\theta^2 \cos\theta - \theta^3 \sin\theta.$

21. At $x = \pi$, $y = \sin\pi = 0$, and the slope $\dfrac{dy}{dx}\Big|_{x=\pi} = \cos x\Big|_{x=\pi} = -1$. Therefore the equation of the tangent line is $y = -(x - \pi) = -x + \pi$. See Figure 3.5.

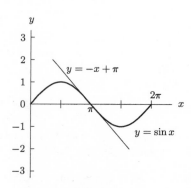

Figure 3.5

25. **(a)** Looking at the graph in Figure 3.6, we see that the maximum is $2600 per month and the minimum is $1400 per month. If $t = 0$ is January 1, then the sales are highest on April 1.

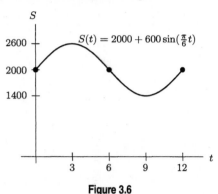

$$S(t) = 2000 + 600 \sin(\frac{\pi}{6}t)$$

Figure 3.6

(b) $S(2)$ is the monthly sales on March 1,

$$S(2) = 2000 + 600 \sin(\frac{\pi}{3})$$
$$= 2000 + 600\sqrt{3}/2 \approx 2519.62 \quad \text{dollars/month}$$

$S'(2)$ is the rate of change of monthly sales on March 1, and since

$$S'(t) = 600[\cos(\frac{\pi}{6}t)](\frac{\pi}{6})$$
$$= 100\pi \cos(\frac{\pi}{6}t),$$

We have,

$$S'(2) = 100\pi \cos(\frac{\pi}{3}) = 50\pi \approx 157.08$$

Solutions for Chapter 3 Review

1. $f'(t) = 24t^3$.

5. $\dfrac{dC}{dq} = 0.08e^{0.08q}$.

9. $\dfrac{d}{dt}e^{(1+3t)^2} = e^{(1+3t)^2}\dfrac{d}{dt}(1+3t)^2 = e^{(1+3t)^2} \cdot 2(1+3t) \cdot 3 = 6(1+3t)e^{(1+3t)^2}$.

13. $f'(x) = 6(3(5x-1)^2) \cdot \dfrac{d}{dx}(5x-1) = 18(5x-1)^2(5) = 90(5x-1)^2$.

17. $\dfrac{dy}{dx} = 2x\ln x + x^2 \cdot \dfrac{1}{x} = x(2\ln x + 1)$.

21. $h'(t) = \dfrac{1}{e^{-t}-t}\left(-e^{-t}-1\right)$.

25. $\dfrac{dy}{dx} = 2x\cos x + x^2(-\sin x) = 2x\cos x - x^2\sin x$.

29. This is a quotient where $u(x) = 1 + e^x$ and $v(x) = 1 - e^{-x}$ so that $q(x) = u(x)/v(x)$.
Using the quotient rule the derivative is

$$q'(x) = \frac{vu' - uv'}{v^2},$$

where $u' = e^x$ and $v' = e^{-x}$. Therefore

$$q'(x) = \frac{(1-e^{-x})e^x - e^{-x}(1+e^x)}{(1-e^{-x})^2} = \frac{e^x - 2 - e^{-x}}{(1-e^{-x})^2}.$$

33. We use the chain rule with $z = g(x) = x^3$ as the inside function and $f(z) = \cos z$ as the outside function. Since $g'(x) = 3x^2$ and $f'(z) = -\sin z$, we have

$$h'(x) = \frac{d}{dx}\left(\cos(x^3)\right) = -\sin z \cdot (3x^2) = -3x^2 \sin(x^3).$$

37. $\frac{dy}{dx} = \frac{1}{3}(\ln 3)3^x - \frac{33}{2}(x^{-\frac{3}{2}}).$

41. $f'(x) = 3x^2 - 8x + 7$, so $f'(0) = 7$, $f'(2) = 3$, and $f'(-1) = 18$.

45. The rate is

$$\frac{dP}{dt} = 35{,}000(\ln 0.98)(0.98)^t.$$

At $t = 45$, this is $35{,}000(\ln 0.98)(0.98)^{45} = -284.9$ people/year. (Note: the negative sign indicates that the population is decreasing.)

49. (a) From the figure, we see $a = 2$. The point with $x = 2$ lies on both the line and the curve. Since

$$y = -3 \cdot 2 + 7 = 1,$$

we have

$$f(a) = 1.$$

Since the slope of the line is -3, we have

$$f'(a) = -3.$$

(b) We use the line to approximate the function, so

$$f(2.1) \approx -3(2.1) + 7 = 0.7.$$

This is an underestimate, because the line is beneath the curve for $x > 2$. Similarly,

$$f(1.98) \approx -3(1.98) + 7 = 1.06.$$

This is an overestimate because the line is above the curve for $x < 2$.

The approximation $f(1.98) \approx 1.06$ is likely to be more accurate because 1.98 is closer to 2 than 2.1 is. Since the graph of $f(x)$ appears to bend away from the line at approximately the same rate on either side of $x = 2$, in this example, the error is larger for points farther from $x = 2$.

53. (a) $V(4) = 25(0.85)^4 = 25(0.522) = 13{,}050$. Thus the value of the car after 4 years is $13,050.
(b) We have a function of the form $f(t) = Ca^t$. We know that such functions have a derivative of the form $(C \ln a) \cdot a^t$. Thus, $V'(t) = 25(0.85)^t \cdot \ln 0.85 = -4.063(0.85)^t$. The units would be the change in value (in thousands of dollars) with respect to time (in years), or thousands of dollars/year.
(c) $V'(4) = -4.063(0.85)^4 = -2.121$. This means that at the end of the fourth year, the value of the car is decreasing by $2121 per year.
(d) $V(t)$ is a positive decreasing function, so that the value of the automobile is positive and decreasing. $V'(t)$ is a negative function whose magnitude is decreasing, meaning the value of the automobile is always dropping, but the yearly loss of value is less as time goes on. The graphs of $V(t)$ and $V'(t)$ confirm that the value of the car decreases with time. What they do not take into account are the *costs* associated with owning the vehicle. At some time, t, it is likely that the yearly costs of owning the vehicle will outweigh its value. At that time, it may no longer be worthwhile to keep the car.

57. The tangent lines to $f(x) = \sin x$ have slope $\frac{d}{dx}(\sin x) = \cos x$. The tangent line at $x = 0$ has slope $f'(0) = \cos 0 = 1$ and goes through the point $(0, 0)$. Consequently, its equation is $y = g(x) = x$. The approximate value of $\sin(\pi/6)$ given by this equation is $g(\pi/6) = \pi/6 \approx 0.524$.

Similarly, the tangent line at $x = \frac{\pi}{3}$ has slope

$$f'\left(\frac{\pi}{3}\right) = \cos\frac{\pi}{3} = \frac{1}{2}$$

and goes through the point $(\pi/3, \sqrt{3}/2)$. Consequently, its equation is

$$y = h(x) = \frac{1}{2}x + \frac{3\sqrt{3} - \pi}{6}.$$

The approximate value of $\sin(\pi/6)$ given by this equation is then

$$h\left(\frac{\pi}{6}\right) = \frac{6\sqrt{3} - \pi}{12} \approx 0.604.$$

The actual value of $\sin(\pi/6)$ is $\frac{1}{2}$, so the approximation from 0 is better than that from $\pi/3$. This is because the slope of the function changes less between $x = 0$ and $x = \pi/6$ than it does between $x = \pi/6$ and $x = \pi/3$. This is illustrated by the following figure.

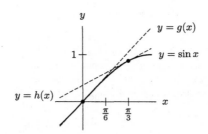

61. From the graphs, we estimate $f(2) \approx 0.3$, $f'(2) \approx 1.1$, $g(2) \approx 1.6$, and $g'(2) \approx -0.5$. By the quotient rule,

$$k'(2) = \frac{f'(2) \cdot g(2) - f(2) \cdot g'(2)}{(g(2))^2} \approx \frac{1.1(1.6) - 0.3(-0.5)}{(1.6)^2} = 0.75.$$

65. (a) We have $p(x) = x^2 - x$. We see that $p'(x) = 2x - 1 < 0$ when $x < \frac{1}{2}$. So p is decreasing when $x < \frac{1}{2}$.

(b) We have $p(x) = x^{1/2} - x$, so

$$p'(x) = \frac{1}{2}x^{-1/2} - 1 < 0$$
$$\frac{1}{2}x^{-1/2} < 1$$
$$x^{-1/2} < 2$$
$$x^{1/2} > \frac{1}{2}$$
$$x > \frac{1}{4}.$$

Thus $p(x)$ is decreasing when $x > \frac{1}{4}$.

(c) We have $p(x) = x^{-1} - x$, so

$$p'(x) = -1x^{-2} - 1 < 0$$
$$-x^{-2} < 1$$
$$x^{-2} > -1,$$

which is always true where x^{-2} is defined since $x^{-2} = 1/x^2$ is always positive. Thus $p(x)$ is decreasing for $x < 0$ and for $x > 0$.

69. All of the functions go through the origin. They will look the same if they have the same tangent line, or equivalently, the same slope at $x = 0$. Therefore for each function we find the derivative and evaluate it at $x = 0$:

$$\begin{array}{lll}
\text{For } y = x, & y' = 1, & \text{so } y'(0) = 1. \\
\text{For } y = \sqrt{x}, & y' = \frac{1}{2\sqrt{x}}, & \text{so } y'(0) \text{ is undefined.} \\
\text{For } y = x^2, & y' = 2x, & \text{so } y'(0) = 0. \\
\text{For } y = x^3 + \frac{1}{2}x^2, & y' = 3x^2 + x, & \text{so } y'(0) = 0. \\
\text{For } y = x^3, & y' = 3x^2, & \text{so } y'(0) = 0. \\
\text{For } y = \ln(x + 1), & y' = \frac{1}{x+1}, & \text{so } y'(0) = 1. \\
\text{For } y = \frac{1}{2}\ln(x^2 + 1), & y' = \frac{x}{x^2+1}, & \text{so } y'(0) = 0. \\
\text{For } y = \sqrt{2x - x^2}, & y' = \frac{1-x}{\sqrt{2x-x^2}}, & \text{so } y'(0) \text{ is undefined.}
\end{array}$$

So near the origin, functions with $y'(0) = 1$ will all be indistinguishable resembling the line $y = x$. These functions are:

$$y = x \quad \text{and} \quad y = \ln(x+1).$$

Functions with $y'(0) = 0$ will be indistinguishable near the origin and resemble the line $y = 0$ (a horizontal line). These functions are:

$$y = x^2, \quad y = x^3 + \frac{1}{2}x^2, \quad y = x^3, \quad \text{and} \quad y = \frac{1}{2}\ln(x^2+1).$$

Functions that have **undefined** derivatives at $x = 0$ look like vertical lines at the origin. These functions are

$$y = \sqrt{x} \quad \text{and} \quad y = \sqrt{2x - x^2}.$$

73. (a) To find the temperature of the yam when it was placed in the oven, we need to evaluate the function at $t = 0$. In this case, the temperature of the yam to begin with equals $350(1 - 0.7e^0) = 350(0.3) = 105°$.

(b) By looking at the function we see that the temperature which the yam is approaching is $350°$. That is, if the yam were left in the oven for a long period of time (i.e. as $t \to \infty$) the temperature would move closer and closer to $350°$ (because $e^{-0.008t}$ would approach zero, and thus $1 - 0.7e^{-0.008t}$ would approach 1). Thus, the temperature of the oven is $350°$.

(c) The yam's temperature will reach $175°$ when $Y(t) = 175$. Thus, we must solve for t:

$$Y(t) = 175$$
$$175 = 350(1 - 0.7e^{-0.008t})$$
$$\frac{175}{350} = 1 - 0.7e^{-0.008t}$$
$$0.7e^{-0.008t} = 0.5$$
$$e^{-0.008t} = 5/7$$
$$\ln e^{-0.008t} = \ln 5/7$$
$$-0.008t = \ln 5/7$$
$$t = \frac{\ln 5/7}{-0.008} \approx 42 \text{ minutes.}$$

Thus the yam's temperature will be $175°$ approximately 42 minutes after it is put into the oven.

(d) The rate at which the temperature is increasing is given by the derivative of the function.

$$Y(t) = 350(1 - 0.7e^{-0.008t}) = 350 - 245e^{-0.008t}.$$

Therefore,

$$Y'(t) = 0 - 245(-0.008e^{-0.008t}) = 1.96e^{-0.008t}.$$

At $t = 20$, the rate of change of the temperature of the yam is given by $Y'(20)$:

$$Y'(20) = 1.96e^{-0.008(20)} = 1.96e^{-.16} = 1.96(0.8521) \approx 1.67 \text{ degrees/minute.}$$

Thus, at $t = 20$ the yam's temperature is increasing by about 1.67 degrees each minute.

Solutions to Problems on Establishing the Derivative Formulas

1. Using the **definition** of the derivative, we have

$$f'(x) = \lim_{h \to 0} \frac{f(x+h) - f(x)}{h}$$
$$= \lim_{h \to 0} \frac{2(x+h) + 1 - (2x+1)}{h}$$
$$= \lim_{h \to 0} \frac{2x + 2h + 1 - 2x - 1}{h}$$
$$= \lim_{h \to 0} \frac{2h}{h}.$$

As long as h is very close to, but not actually equal to, zero we can say that $\lim_{h \to 0} \frac{2h}{h} = 2$, and thus conclude that $f'(x) = 2$.

5. The definition of the derivative states that

$$f'(x) = \lim_{h \to 0} \frac{f(x+h) - f(x)}{h}.$$

Using this definition, we have

$$f'(x) = \lim_{h \to 0} \frac{4(x+h)^2 + 1 - (4x^2 + 1)}{h}$$

$$= \lim_{h \to 0} \frac{4x^2 + 8xh + 4h^2 + 1 - 4x^2 - 1}{h}$$

$$= \lim_{h \to 0} \frac{8xh + 4h^2}{h}$$

$$= \lim_{h \to 0} \frac{h(8x + 4h)}{h}.$$

As long as h approaches, but does not equal, zero we can cancel h in the numerator and denominator. The derivative now becomes

$$\lim_{h \to 0} (8x + 4h) = 8x.$$

Thus, $f'(x) = 6x$ as we stated above.

9. Since $f(x) = C$ for all x, we have $f(x+h) = C$. Using the definition of the derivative, we have

$$f'(x) = \lim_{h \to 0} \frac{f(x+h) - f(x)}{h}$$

$$= \lim_{h \to 0} \frac{C - C}{h}$$

$$= \lim_{h \to 0} \frac{0}{h}.$$

As h gets very close to zero without actually equaling zero, we have $0/h = 0$, so

$$f'(x) = \lim_{h \to 0} (0) = 0.$$

Solutions to Practice Problems on Differentiation

1. $f'(t) = 2t + 4t^3$

5. $f'(x) = -2x^{-3} + 5\left(\frac{1}{2}x^{-1/2}\right) = \dfrac{-2}{x^3} + \dfrac{5}{2\sqrt{x}}$

9. $D'(p) = 2pe^{p^2} + 10p$

13. $s'(t) = \dfrac{16}{2t + 1}$

17. $C'(q) = 3(2q + 1)^2 \cdot 2 = 6(2q + 1)^2$

21. $y' = 2x \ln(2x + 1) + \dfrac{2x^2}{2x + 1}$

25. $g'(t) = 15 \cos(5t)$

29. $y' = 17 + 12x^{-1/2}$.

33. Either notice that $f(x) = \dfrac{x^2 + 3x + 2}{x + 1}$ can be written as $f(x) = \dfrac{(x + 2)(x + 1)}{x + 1}$ which reduces to $f(x) = x + 2$, giving $f'(x) = 1$, or use the quotient rule which gives

$$f'(x) = \frac{(x + 1)(2x + 3) - (x^2 + 3x + 2)}{(x + 1)^2}$$

$$= \frac{2x^2 + 5x + 3 - x^2 - 3x - 2}{(x+1)^2}$$
$$= \frac{x^2 + 2x + 1}{(x+1)^2}$$
$$= \frac{(x+1)^2}{(x+1)^2}$$
$$= 1.$$

37. $q'(r) = \dfrac{3(5r+2) - 3r(5)}{(5r+2)^2} = \dfrac{15r + 6 - 15r}{(5r+2)^2} = \dfrac{6}{(5r+2)^2}$

41. $h'(w) = 5(w^4 - 2w)^4(4w^3 - 2)$

45. $h'(w) = 6w^{-4} + \dfrac{3}{2}w^{-1/2}$

49. Using the chain rule, $g'(\theta) = (\cos\theta)e^{\sin\theta}$.

53. $h'(r) = \dfrac{d}{dr}\left(\dfrac{r^2}{2r+1}\right) = \dfrac{(2r)(2r+1) - 2r^2}{(2r+1)^2} = \dfrac{2r(r+1)}{(2r+1)^2}.$

57. $f'(x) = \dfrac{3x^2}{9}(3\ln x - 1) + \dfrac{x^3}{9}\left(\dfrac{3}{x}\right) = x^2 \ln x - \dfrac{x^2}{3} + \dfrac{x^2}{3} = x^2 \ln x$

61. Using the quotient rule gives

$$w'(r) = \frac{2ar(b+r^3) - 3r^2(ar^2)}{(b+r^3)^2}$$
$$= \frac{2abr - ar^4}{(b+r^3)^2}.$$

CHAPTER FOUR

Solutions for Section 4.1

1. We find a critical point by noting where $f'(t) = 0$ or f' is undefined. Since the curve is smooth throughout, f' is always defined, so we look for where $f'(t) = 0$, or equivalently where the tangent line to the graph is horizontal. These points are shown in Figure 4.1.

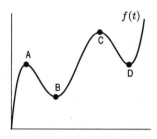

Figure 4.1

As we can see, there are four labeled critical points. Critical point A is a local maximum because points near it are all lower; similarly, point B is a local minimum, point C is a local maximum, and point D is a local minimum.

5. There was a critical point after the first eighteen hours when temperature was at its highest point, a local maximum for the temperature function.

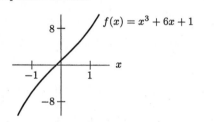

Figure 4.2

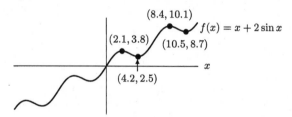

Figure 4.3

9. The graph of f in Figure 4.2 appears to be increasing for all x, with no critical points. Since $f'(x) = 3x^2 + 6$ and $x^2 \geq 0$ for all x, we have $f'(x) > 0$ for all x. That explains why f is increasing for all x.

13. The graph of f in Figure 4.3 looks like a climbing sine curve, alternately increasing and decreasing, with more time spent increasing than decreasing. Here $f'(x) = 1 + 2\cos x$, so $f'(x) = 0$ when $\cos x = -1/2$; this occurs when

$$x = \pm\frac{2\pi}{3}, \pm\frac{4\pi}{3}, \pm\frac{8\pi}{3}, \pm\frac{10\pi}{3}, \pm\frac{14\pi}{3}, \pm\frac{16\pi}{3}\ldots$$

Since $f'(x)$ changes sign at each of these values, the graph of f must alternate increasing/decreasing. However, the distance between values of x for critical points alternates between $(2\pi)/3$ and $(4\pi)/3$, with $f'(x) > 0$ on the intervals of length $(4\pi)/3$. For example, $f'(x) > 0$ on the interval $(4\pi)/3 < x < (8\pi)/3$. As a result, f is increasing on the intervals of length $(4\pi/3)$ and decreasing on the intervals of length $(2\pi/3)$.

17. The critical points of f occur where f' is zero. These two points are indicated in the figure below.

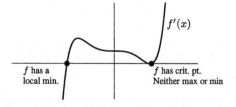

Note that the point labeled as a local minimum of f is not a critical point of f'.

21. Since $f'(x) = 4x^3 - 12x^2 + 8$, we see that $f'(1) = 0$, as we expected. We apply the second derivative test to $f''(x) = 12x^2 - 24x$. Since $f''(1) = -12 < 0$, the graph is concave down at the critical point $x = 1$, making it a local maximum.

25.

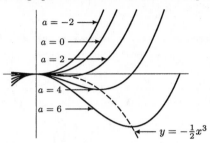

To solve for the critical points, we set $\frac{dy}{dx} = 0$. Since $\frac{d}{dx}\left(x^3 - ax^2\right) = 3x^2 - 2ax$, we want $3x^2 - 2ax = 0$, so $x = 0$ or $x = \frac{2}{3}a$. At $x = 0$, we have $y = 0$. This first critical point is independent of a and lies on the curve $y = -\frac{1}{2}x^3$. At $x = \frac{2}{3}a$, we calculate $y = -\frac{4}{27}a^3 = -\frac{1}{2}\left(\frac{2}{3}a\right)^3$. Thus the second critical point also lies on the curve $y = -\frac{1}{2}x^3$.

29. (a) In Figure 4.4, we see that $f(\theta) = \theta - \sin\theta$ has a zero at $\theta = 0$. To see if it has any other zeros near the origin, we use our calculator to zoom in. (See Figure 4.5.) No extra root seems to appear no matter how close to the origin we zoom. However, zooming can never tell you for sure that there is not a root that you have not found yet.

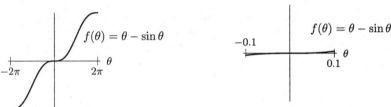

Figure 4.4: Graph of $f(\theta)$

Figure 4.5: Graph of $f(\theta)$ Zoomed In

(b) Using the derivative, $f'(\theta) = 1 - \cos\theta$, we can argue that there is no other zero. Since $\cos\theta < 1$ for $0 < \theta \le 1$, we know $f'(\theta) > 0$ for $0 < \theta \le 1$. Thus, f increases for $0 < \theta \le 1$. Consequently, we conclude that the only zero of f is the one at the origin. If f had another zero at x_0, with $x_0 > 0$, then f would have to "turn around", and recross the x-axis at x_0. But if this were the case, f' would be nonpositive somewhere, which we know is not the case.

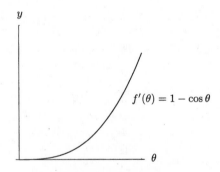

Figure 4.6: Graph of $f'(\theta)$

Solutions for Section 4.2

1. We find an inflection point by noting where the concavity changes. Such points are shown in Figure 4.7. There are three inflection points.

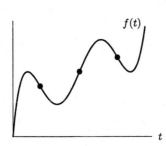

Figure 4.7

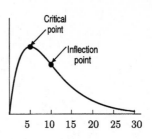

Figure 4.8

5. One possible answer is shown in Figure 4.8.

9. (a) Critical point.
(b) Inflection point.

13. $f'(x) = 6x^2 + 6x - 36$. To find critical points, we set $f'(x) = 0$. Then

$$6(x^2 + x - 6) = 6(x + 3)(x - 2) = 0.$$

Therefore, the critical points of f are $x = -3$ and $x = 2$. To find the inflection points of $f(x)$ we look for the points at which $f''(x)$ goes from negative to positive or vice-versa. Since $f''(x) = 12x + 6$, $x = -1/2$ is an inflection point. From Figure 4.9, we see the critical point $x = -3$ is a local maximum and the critical point $x = 2$ is a local minimum.

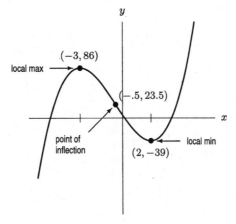

Figure 4.9

17. The derivative of $f(x)$ is $f'(x) = 4x^3 - 16x$. The critical points of $f(x)$ are points at which $f'(x) = 0$. Factoring, we get

$$4x^3 - 16x = 0$$
$$4x(x^2 - 4) = 0$$
$$4x(x - 2)(x + 2) = 0$$

So the critical points of f will be $x = 0, x = 2$, and $x = -2$.

To find the inflection points of f we look for the points at which $f''(x)$ changes sign. At any such point $f''(x)$ is either zero or undefined. Since $f''(x) = 12x^2 - 16$ our candidate points are $x = \pm 2/\sqrt{3}$. At both of these points $f''(x)$ changes sign, so both of these points are inflection points.

From Figure 4.10, we see that the critical points $x = -2$ and $x = 2$ are local minima and the critical point $x = 0$ is a local maximum.

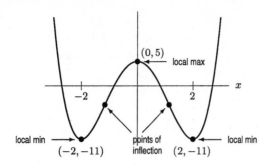

Figure 4.10

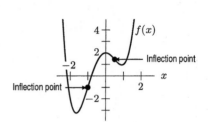

Figure 4.11

21. From the graph of $f(x)$ in the figure below, we see that the function must have two inflection points. We calculate $f'(x) = 4x^3 + 3x^2 - 6x$, and $f''(x) = 12x^2 + 6x - 6$. Solving $f''(x) = 0$ we find that:

$$x_1 = -1 \quad \text{and} \quad x_2 = \frac{1}{2}.$$

Since $f''(x) > 0$ for $x < x_1$, $f''(x) < 0$ for $x_1 < x < x_2$, and $f''(x) > 0$ for $x_2 < x$, it follows that both points are inflection points. See Figure 4.11.

25. See Figure 4.12.

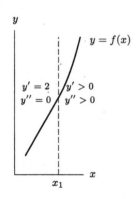

Figure 4.12

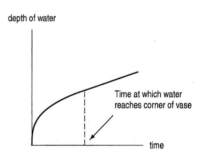

Figure 4.13

29. See Figure 4.13.

33. **(a)** An inflection point occurs whenever the concavity of $f(x)$ changes. If the graph shown is that of $f(x)$, then an inflection point will occur whenever its concavity changes, or equivalently when the tangent line moves from above the curve to below or vice-versa. Such points are shown in Figure 4.14.

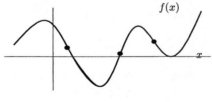

Figure 4.14

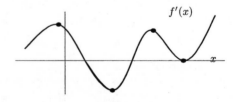

Figure 4.15

(b) To find inflection points of the function f we must find points where f'' changes sign. However, because f'' is the derivative of f', any point where f'' changes sign will be a local maximum or minimum on the graph of f'. Such points are shown in Figure 4.15.

(c) The inflection points of f are the points where f'' changes sign. If the graph shown is that of $f''(x)$, then we are looking for where the given graph passes from above the x-axis to below, or vice versa. Such points are shown in Figure 4.16:

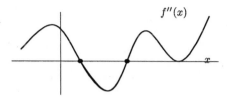

$f''(x)$

Figure 4.16

Solutions for Section 4.3

1. See Figure 4.17.

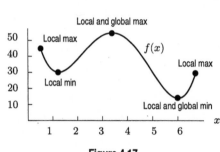

Local and global max

Local max

$f(x)$

Local max

Local min

Local and global min

Figure 4.17

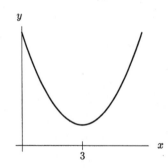

y

x

3

Figure 4.18

5. See Figure 4.18.

9. True. If the maximum is not at an endpoint, then it must be at critical point of f. But $x = 0$ is the only critical point of $f(x) = x^2$ and it gives a minimum, not a maximum.

13. See Figure 4.19.

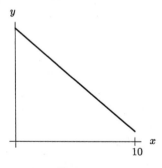

y

x

10

Figure 4.19

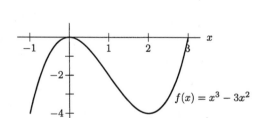

$f(x) = x^3 - 3x^2$

Figure 4.20

17. (a) Differentiating $f(x) = x^3 - 3x^2$ produces $f'(x) = 3x^2 - 6x$. A second differentiation produces $f''(x) = 6x - 6$.
  (b) $f'(x)$ is defined for all x and $f'(x) = 0$ when $x = 0, 2$. Thus 0 and 2 are the critical points of f.
  (c) $f''(x)$ is defined for all x and $f''(x) = 0$ when $x = 1$. When $x < 1$, $f''(x) < 0$ and when $x > 1$, $f''(x) > 0$. Thus the concavity of the graph of f changes at $x = 1$. Hence $x = 1$ is an inflection point.
  (d) $f(-1) = -4$, $f(0) = 0$, $f(2) = -4$, $f(3) = 0$. So f has a local maximum at $x = 0$, a local minimum at $x = 2$, global maxima at $x = 0$ and $x = 3$, and global minima at $x = -1$ and $x = 2$.
  (e) Plotting the function $f(x)$ for $-1 \le x \le 3$ gives the graph shown in Figure 4.20.

21. (a) Differentiating $f(x) = e^{-x} \sin x$ produces $f'(x) = -e^{-x} \sin x + e^{-x} \cos x$. A second differentiation produces $f''(x) = -2e^{-x} \cos x$.

(b) $f'(x)$ is defined for all x and $f'(x) = 0$ when $x = \pi/4$ and when $x = 5\pi/4$. Thus $\pi/4$ and $5\pi/4$ are the critical points of f.

(c) $f''(x)$ is defined for all x and $f''(x) = 0$ when $x = \pi/2, 3\pi/2$. Since the concavity of f changes at both of these points they are both inflection points.

(d) $f(0) = 0$, $f(2\pi) = 0$, $f(\pi/4) = e^{-\pi/4} \sin \pi/4$ and $f(5\pi/4) = e^{-5\pi/4} \sin(5\pi/4)$. So f has a global maximum at $x = \pi/4$ and a global minimum at $x = 5\pi/4$.

(e) Plotting the function $f(x)$ for $0 \le x \le 2\pi$ gives the graph shown in Figure 4.21.

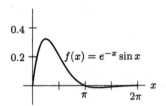

Figure 4.21

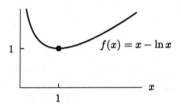

Figure 4.22

25. Differentiating gives

$$f'(x) = 1 - \frac{1}{x},$$

so the critical points satisfy

$$1 - \frac{1}{x} = 0$$
$$\frac{1}{x} = 1$$
$$x = 1.$$

Since f' is negative for $0 < x < 1$ and f' is positive for $x > 1$, there is a local minimum at $x = 1$.

Since $f(x) \to \infty$ as $x \to 0^+$ and as $x \to \infty$, the local minimum at $x = 1$ is a global minimum; there is no global maximum. See Figure 4.22. Thus, the global minimum is $f(1) = 1$.

29. The speed is given for r in the interval $0 \le r \le R$. We have $v(r) = a(R - r)r^2 = aRr^2 - ar^3$, and $v'(r) = 2aRr - 3ar^2 = 2ar(R - \frac{3}{2}r)$, which is zero if $r = \frac{2}{3}R$, or if $r = 0$, and so $v(r)$ has critical points there.

Since $r = \frac{2}{3}R$ is the only critical point in the interval $0 < r < R$ and $v(0) = v(R) = 0$, we know that $r = \frac{2}{3}R$ is the global maximum.

33. Consider the rectangle of sides x and y shown in the figure below.

The total area is $xy = 3000$, so $y = 3000/x$. Suppose the left and right edges and the lower edge have the shrubs and the top edge has the fencing. The total cost is

$$C = 25(x + 2y) + 10(x)$$
$$= 35x + 50y.$$

Since $y = 3000/x$, this reduces to

$$C(x) = 35x + 50(3000/x) = 35x + 150{,}000/x.$$

Therefore, $C'(x) = 35 - 150{,}000/x^2$. We set this to 0 to find the critical points:

$$35 - \frac{150{,}000}{x^2} = 0$$
$$\frac{150{,}000}{x^2} = 35$$
$$x^2 = 4285.71$$
$$x \approx 65.5 \text{ ft}$$

so that

$$y = 3000/x \approx 45.8 \text{ ft.}$$

Since $C(x) \to \infty$ as $x \to 0^+$ and $x \to \infty$, $x = 65.5$ is a minimum. The minimum total cost is then

$$C(65.5) \approx \$4583.$$

37. Rewriting the expression for I using the properties of logs gives

$$I = 192(\ln S - \ln 762) - S + 763.$$

Differentiating with respect to S gives

$$\frac{dI}{dS} = \frac{192}{S} - 1.$$

At a critical point

$$\frac{192}{S} - 1 = 0$$
$$S = 192.$$

Since

$$\frac{d^2 I}{dS^2} = -\frac{192}{S^2},$$

we see that if $S = 192$, we have $d^2 I/dS^2 < 0$, so $S = 192$ is a local maximum. From Figure 4.23, we see that it is a global maximum. The maximum possible number of infected children is therefore

$$I = 192 \ln\left(\frac{192}{762}\right) - 192 + 763 = 306 \text{ children.}$$

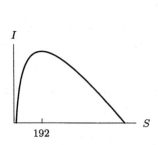

Figure 4.23

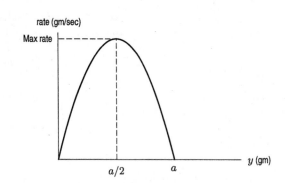

Figure 4.24

41. (a) If we expect the rate to be nonnegative, then we must have $0 \le y \le a$. See Figure 4.24.
   (b) The maximum value of the rate occurs at $y = a/2$, as can be seen from Figure 4.24, or by setting

$$\frac{d}{dy}(\text{rate}) = 0$$
$$\frac{d}{dy}(\text{rate}) = \frac{d}{dy}(kay - ky^2) = ka - 2ky = 0$$
$$y = \frac{a}{2}.$$

From the graph, we see that $y = a/2$ gives the global maximum.

45. (a) To maximize benefit (surviving young), we pick 10, because that's the highest point of the benefit graph.

(b) To optimize the vertical distance between the curves, we can either do it by inspection or note that the slopes of the two curves will be the same where the difference is maximized. Either way, one gets approximately 9.

49. The triangle in Figure 4.25 has area, A, given by

$$A = \frac{1}{2}xy = \frac{x}{2}e^{-x/3}.$$

At a critical point,

$$\frac{dA}{dx} = \frac{1}{2}e^{-x/3} - \frac{x}{6}e^{-x/3} = 0$$

$$\frac{1}{6}e^{-x/3}(3-x) = 0$$

$$x = 3.$$

Substituting the critical point and the endpoints into the formula for the area gives:

For $x = 1$, we have $A = \frac{1}{2}e^{-1/3} = 0.358$

For $x = 3$, we have $A = \frac{3}{2}e^{-1} = 0.552$

For $x = 5$, we have $A = \frac{5}{2}e^{-5/3} = 0.472$

Thus, the maximum area is 0.552 and the minimum area is 0.358.

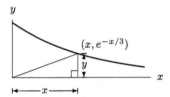

Figure 4.25

Solutions for Section 4.4

1. (a) Profit is maximized when $R(q) - C(q)$ is as large as possible. This occurs at $q = 2500$, where profit $= 7500 - 5500 = \$2000$.

(b) We see that $R(q) = 3q$ and so the price is $p = 3$, or \$3 per unit.

(c) Since $C(0) = 3000$, the fixed costs are \$3000.

5. (a) The profit earned by the 51st is the revenue earned by the 51st item minus the cost of producing the 51st item. This can be approximated by

$$\pi'(50) = R'(50) - C'(50) = 84 - 75 = \$9.$$

Thus the profit earned from the 51st item will be approximately \$9.

(b) The profit earned by the 91st item will be the revenue earned by the 91st item minus the cost of producing the 91st item. This can be approximated by

$$\pi'(90) = R'(90) - C'(90) = 68 - 71 = -\$3.$$

Thus, approximately three dollars are lost in the production of the 91st item.

(c) If $R'(78) > C'(78)$, production of a 79th item would increase profit. If $R'(78) < C'(78)$, production of one less item would increase profit. Since profit is maximized at $q = 78$, we must have

$$C'(78) = R'(78).$$

9. (a) Profit $= \pi = R - C$; profit is maximized when the slopes of the two graphs are equal, at around $q = 350$. See Figure 4.26.

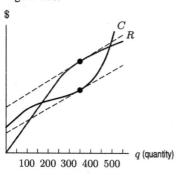

| Figure 4.26 | Figure 4.27 |

(b) The graphs of MR and MC are the derivatives of the graphs of R and C. Both R and C are increasing everywhere, so MR and MC are everywhere positive. The cost function is concave down and then concave up, so MC is decreasing and then increasing. The revenue function is linear and then concave down, so MR is constant and then decreasing. See Figure 4.27.

13. The profit is maximized at the point where the difference between revenue and cost is greatest. Thus the profit is maximized at approximately $q = 4000$.

17. We first need to find an expression for $R(q)$, or revenue in terms of quantity sold. We know that $R(q) = pq$, where p is the price of one item. Here $p = 45 - 0.01q$, so we make the substitution

$$R(q) = (45 - .01q)q = 45q - 0.01q^2.$$

This is the function we want to maximize. Finding the derivative and setting it equal to 0 yields

$$R'(q) = 0$$
$$45 - 0.02q = 0$$
$$0.02q = 45 \text{ so}$$
$$q = 2250.$$

Is this a maximum?

$$R'(q) > 0 \text{ for } q < 2250 \text{ and}$$
$$R'(q) < 0 \text{ for } q > 2250.$$

So we conclude that $R(q)$ has a local maximum at $q = 2250$. Testing $q = 0$, the only endpoint, $R(0) = 0$, which is less than $R(2250) = \$50,625$. So we conclude that revenue is maximized at $q = 2250$. The price of each item at this production level is

$$p = 45 - .01(2250) = \$22.50$$

and total revenue is

$$pq = \$22.50(2250) = \$50,625,$$

which agrees with the above answer.

21. We first need to find an expression for revenue in terms of price. At a price of \$8, 1500 tickets are sold. For each \$1 above \$8, 75 fewer tickets are sold. This suggests the following formula for q, the quantity sold for any price p.

$$q = 1500 - 75(p - 8)$$
$$= 1500 - 75p + 600$$
$$= 2100 - 75p.$$

We know that $R = pq$, so substitution yields

$$R(p) = p(2100 - 75p) = 2100p - 75p^2$$

To maximize revenue, we find the derivative of $R(p)$ and set it equal to 0.

$$R'(p) = 2100 - 150p = 0$$
$$150p = 2100$$

so $p = \frac{2100}{150} = 14$. Does $R(p)$ have a maximum at $p = 14$? Using the first derivative test,

$$R'(p) > 0 \text{ if } p < 14 \text{ and}$$
$$R'(p) < 0 \text{ if } p > 14.$$

So $R(p)$ has a local maximum at $p = 14$. Since this is the only critical point for $p \geq 0$, it must be a global maximum. So we conclude that revenue is maximized when the price is $14.

25. Let x equal the number of chairs ordered in excess of 300, so $0 \leq x \leq 100$.

$$\text{Revenue} = R = (90 - 0.25x)(300 + x)$$
$$= 27,000 - 75x + 90x - 0.25x^2 = 27,000 + 15x - 0.25x^2$$

At a critical point $dR/dx = 0$. Since $dR/dx = 15 - 0.5x$, we have $x = 30$, and the maximum revenue is $27,225$ since the graph of R is a parabola which opens downward. The minimum is $0 (when no chairs are sold).

29. For each month,

$$\text{Profit} = \text{Revenue} - \text{Cost}$$
$$\pi = pq - wL = pcK^\alpha L^\beta - wL$$

The variable on the right is L, so at the maximum

$$\frac{d\pi}{dL} = \beta p c K^\alpha L^{\beta-1} - w = 0$$

Now $\beta - 1$ is negative, since $0 < \beta < 1$, so $1 - \beta$ is positive and we can write

$$\frac{\beta p c K^\alpha}{L^{1-\beta}} = w$$

giving

$$L = \left(\frac{\beta p c K^\alpha}{w}\right)^{\frac{1}{1-\beta}}$$

Since $\beta - 1$ is negative, when L is just above 0, the quantity $L^{\beta-1}$ is huge and positive, so $d\pi/dL > 0$. When L is large, $L^{\beta-1}$ is small, so $d\pi/dL < 0$. Thus the value of L we have found gives a global maximum, since it is the only critical point.

Solutions for Section 4.5

1. (a) The average cost of quantity q is given by the formula $C(q)/q$. So average cost at $q = 10,000$ is given by $C(10,000)/10,000$. From the graph, we see that $C(10,000) \approx 16,000$, so $a(q) \approx \frac{16,000}{10,000} \approx \$1.60$ per unit. The economic interpretation of this is that $1.60 is each unit's share of the total cost of producing 10,000 units.

(b) To interpret this graphically, note that $a(q) = \frac{C(q)}{q} = \frac{C(q)-0}{q-0}$. This is exactly the formula for the slope of a line from the origin to a point $(q, C(q))$ on the curve. So $a(10,000)$ is the slope of a line connecting $(0,0)$ to $(10,000, C(10,000))$. Such a line is shown below in Figure 4.28.

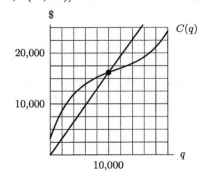

Figure 4.28

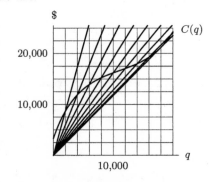

Figure 4.29

(c) We know that $a(q)$ is minimized where $a(q) = C'(q)$. Using the graphical interpretations from parts (b) and (c), this is equivalent to saying that the tangent line has the same slope as the line connecting the point on the curve to the origin. Since these two lines share a point, specifically the point $(q, C(q))$ on the curve, and have the same slope, they are in fact the same line. So $a(q)$ is minimized where the line passing from $(q, C(q))$ to the origin is also tangent to the curve. To find such points, a variety of lines passing through the origin and the curve are shown in Figure 4.29.

From this plot, we see that the line with the desired properties intersects the curve at $q \approx 18,000$. So $q \approx 18,000$ units minimizes $a(q)$.

5. The graph of the average cost function is shown in Figure 4.30.

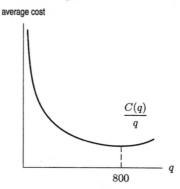

Figure 4.30

9. **(a)** $a(q) = C(q)/q$, so $C(q) = 0.01q^3 - 0.6q^2 + 13q$.

(b) Taking the derivative of $C(q)$ gives an expression for the marginal cost:

$$C'(q) = MC(q) = 0.03q^2 - 1.2q + 13.$$

To find the smallest MC we take its derivative and find the value of q that makes it zero. So: $MC'(q) = 0.06q - 1.2 = 0$ when $q = 1.2/0.06 = 20$. This value of q must give a minimum because the graph of $MC(q)$ is a parabola opening upward. Therefore the minimum marginal cost is $MC(20) = 1$. So the marginal cost is at a minimum when the additional cost per item is \$1.

(c) $a'(q) = 0.02q - 0.6$

Setting $a'(q) = 0$ and solving for q gives $q = 30$ as the quantity at which the average is minimized, since the graph of a is a parabola which opens upward. The minimum average cost is $a(30) = 4$ dollars per item.

(d) The marginal cost at $q = 30$ is $MC(30) = 0.03(30)^2 - 1.2(30) + 13 = 4$. This is the same as the average cost at this quantity. Note that since $a(q) = C(q)/q$, we have $a'(q) = (qC'(q) - C(q))/q^2$. At a critical point, q_0, of $a(q)$, we have

$$0 = a'(q_0) = \frac{q_0 C'(q_0) - C(q_0)}{q_0^2},$$

so $C'(q_0) = C(q_0)/q_0 = a(q_0)$. Therefore $C'(30) = a(30) = 4$ dollars per item.

Another way to see why the marginal cost at $q = 30$ must equal the minimum average cost $a(30) = 4$ is to view $C'(30)$ as the approximate cost of producing the 30th or 31st good. If $C'(30) < a(30)$, then producing the 31st good would lower the average cost, i.e. $a(31) < a(30)$. If $C'(30) > a(30)$, then producing the 30th good would raise the average cost, i.e. $a(30) > a(29)$. Since $a(30)$ is the global minimum, we must have $C'(30) = a(30)$.

13. It is interesting to note that to draw a graph of $C'(q)$ for this problem, you never have to know what $C(q)$ looks like, although you *could* draw a graph of $C(q)$ if you wanted to. By the definition of average cost, we know that $C(q) = q \cdot a(q)$. Using the product rule we get that $C'(q) = a(q) + q \cdot a'(q)$.

We are given a graph of $a(q)$ which is linear, so $a(q) = b + mq$, where $b = a(0)$ is the y-intercept and m is the slope. Therefore

$$C'(q) = a(q) + q \cdot a'(q) = b + mq + q \cdot m$$
$$= b + 2mq.$$

In other words, $C'(q)$ is also linear, and it has twice the slope and the same y-intercept as $a(q)$.

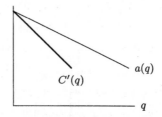

Solutions for Section 4.6

1. The effect on demand is approximately E times the change in price. A price increase causes a decrease in demand and a price decrease causes an increase in demand.

   (a) Demand will decrease by about $0.5(3\%) = 1.5\%$.

   (b) Demand will increase by about $0.5(3\%) = 1.5\%$.

5. Demand for high-definition TV's will be elastic, since it is not a necessary item. If the prices are too high, people will not choose to buy them, so price changes will cause relatively large demand changes.

9. The elasticity of demand is given by

$$E = \left| \frac{p}{q} \cdot \frac{dq}{dp} \right|.$$

To evaluate dq/dp we solve $p = 90 - 10q$ for q, so $q = (90 - p)/10$ and hence

$$\frac{dq}{dp} = -\frac{1}{10}.$$

When $p = 50$, we find $q = 4$, $dq/dp = -1/10$ so

$$E = \left| \frac{50}{4} \cdot \left(-\frac{1}{10} \right) \right| = 1.25.$$

Since the elasticity is

$$E = \frac{\text{Percent change in demand}}{\text{Percent change in price}}$$

when the price increases by 2% the percent change in demand is given by

$$\text{Percent change in demand} = E \cdot \text{Percent change in price}$$
$$= 2 \cdot (1.25) = 2.5.$$

Therefore, the percentage change in demand is 2.5%. Since $dq/dp < 0$ this corresponds to a 2.5% fall in demand.

13. **(a)** At a price of \$2/pound, the quantity sold is

$$q = 5000 - 10(2)^2 = 5000 - 40 = 4960$$

so the total revenue is

$$R = pq = 2 \cdot 4960 = \$9{,}920$$

   (b) We know that $R = pq$, and that $q = 5000 - 10p^2$, so we can substitute for q to find $R(p)$

$$R(p) = p(5000 - 10p^2) = 5000p - 10p^3$$

To find the price that maximizes revenue we take the derivative and set it equal to 0.

$$R'(p) = 0$$
$$5000 - 30p^2 = 0$$
$$30p^2 = 5000$$
$$p^2 = 166.67$$
$$p = \pm 12.91$$

We disregard the negative answer, so $p = 12.91$ is the only critical point. Is it the maximum? We use the first derivative test.

$$R'(p) > 0 \ \text{ if } \ p < 12.91 \ \text{ and }$$
$$R'(p) < 0 \ \text{ if } \ p > 12.91$$

So $R(p)$ has a local maximum at $p = 12.91$. We also test the function at $p = 0$, which is the only endpoint.

$$R(0) = 5000(0) - 10(0)^3 = 0$$

$$R(12.91) = 5000(12.91) - 10(12.91)^3 = 64{,}550 - 21{,}516.85 = \$43{,}033.15$$

So we conclude that revenue is maximized at price of \$12.91/pound.

(c) At a price of $12.91/pound the quantity sold is

$$q = 5000 - 10(12.91)^2 = 5000 - 1666.68 = 3333.32$$

so the total revenue is

$$R = pq = (3333.32)(12.91) = \$43,033.16$$

which agrees with part (b).

(d)

$$E = \left| \frac{p}{q} \cdot \frac{dq}{dp} \right| = \left| \frac{p}{q} \cdot \frac{d}{dp}(5000 - 10p^2) \right| = \left| \frac{p}{q} \cdot (-20p) \right| = \frac{20p^2}{q}$$

Substituting $p = 12.91$ and $q = 3333.32$ yields

$$E = \frac{20(12.91)^2}{3333.32} = \frac{3333.36}{3333.32} \approx 1$$

which agrees with the result that maximum revenue occurs when $E = 1$.

17. Demand is inelastic at all prices. No matter what the price is, you can increase revenue by raising the price, so there is no actual price for which your revenue is maximized. This is not a realistic example, but it is mathematically possible. It would correspond, for instance, to the demand equation $q = 1/\sqrt{p}$, which gives revenue $R = pq = \sqrt{p}$ which is increasing for all prices $p > 0$.

21. Since marginal revenue equals dR/dq and $R = pq$, we have, using the product rule,

$$\frac{dR}{dq} = \frac{d(pq)}{dq} = p \cdot 1 + \frac{dp}{dq} \cdot q = p\left(1 + \frac{q}{p} \cdot \frac{dp}{dq}\right) = p\left(1 - \frac{1}{-\frac{p}{q} \cdot \frac{dq}{dp}}\right) = p\left(1 - \frac{1}{E}\right).$$

25. The approximation $E_{\text{income}} \approx \left| \frac{\Delta q/q}{\Delta I/I} \right|$ shows that the income elasticity measures the ratio of the fractional change in quantity of the product demanded to the fractional change in the income of the consumer. Thus, for example, a 1% increase in income will translate into an $E_{\text{income}}\%$ increase in the quantity purchased. After an increase in income, the consumer will tend to buy more. The income elasticity measures the strength of this tendency.

Solutions for Section 4.7

1. (a) As t gets very very large, $e^{-0.08t} \to 0$ and the function becomes $P \approx 40/1$. Thus, this model implies that when t is very large, the population is 40 billion.

   (b) A graph of P against t is shown in Figure 4.31.

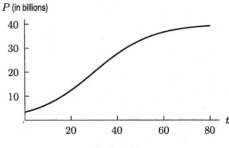

   P (in billions)

   Figure 4.31

   (c) We are asked to find the time t such that $P(t) = 20$. Solving we get

$$20 = P(t) = \frac{40}{1 + 11e^{-0.08t}}$$

$$1 + 11e^{-0.08t} = \frac{40}{20} = 2$$

$$11e^{-0.08t} = 1$$

$$e^{-0.08t} = \frac{1}{11}$$

$$\ln e^{-0.08t} = \ln \frac{1}{11}$$

$$-0.08t \approx -2.4$$

$$t \approx \frac{-2.4}{-0.08} = 30$$

Thus 30 years from 1990 (the year 2020) the population of the world should be 20 billion.

We are asked to find the time t such that $P(t) = 39.9$. Solving we get

$$39.9 = P(t) = \frac{40}{1 + 11e^{-0.08t}}$$

$$1 + 11e^{-0.08t} = \frac{40}{39.9} = 1.00251$$

$$11e^{-0.08t} = 0.00251$$

$$e^{-0.08t} = \frac{0.00251}{11}$$

$$\ln e^{-0.08t} = \ln \frac{0.00251}{11}$$

$$-0.08t \approx -8.39$$

$$t \approx \frac{-8.39}{-0.08} \approx 105$$

Thus 105 years from 1990 (the year 2095) the population of the world should be 39.9 billion.

5. Sales of a new product could very well follow a logistic curve. At first, sales will grow exponentially as more and more people hear of the product and decide to buy it. Eventually, though, everyone will know about the product and while people may still buy it, not as many will (sales will slow down). Eventually, it's possible that everyone who would buy the product already has, at which point sales will stop. It behooves the seller to notice the point of diminishing returns so they don't make more of their product than people will want to buy.

9. (a) At $t = 0$, which corresponds to 1935, we have

$$P = \frac{1}{1 + 3e^{-0.0275(0)}} = 0.25$$

showing that 25% of the land was in use in 1935.

(b) This model predicts that as t gets very large, P approaches 1. That is, the model predicts that in the long run, all the land will be used for farming.

(c) To solve this graphically, enter the function into a graphing calculator and trace the resulting curve until it reaches a height of 0.5, which occurs when $t = 39.9 \approx 40$. Since $t = 0$ corresponds to 1935, $t = 40$ corresponds to $1935 + 40 = 1975$. According to this model, the Tojolobal were using half their land in 1975. See Figure 4.32.

(d) The point of diminishing returns occurs when $P = L/2$ or at one-half the carrying capacity. In this case, $P = 1/2$ in 1975, as shown in part (c).

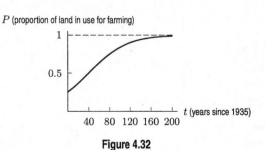

Figure 4.32

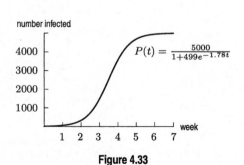

Figure 4.33

13. (a) We use $k = 1.78$ as a rough approximation. We let $L = 5000$ since the problem tells us that 5000 people eventually get the virus. This means the limiting value is 5000.

(b) We know that

$$P(t) = \frac{5000}{1 + Ce^{-1.78t}} \quad \text{and} \quad P(0) = 10$$

so

$$10 = \frac{5000}{1 + Ce^0} = \frac{5000}{1 + C}$$
$$10(1 + C) = 5000$$
$$1 + C = 500$$
$$C = 499.$$

(c) We have $P(t) = \dfrac{5000}{1 + 499e^{-1.78t}}$. This function is graphed in Figure 4.33.

(d) The point of diminishing returns appears to be at the point $(3.5, 2500)$; that is, after 3 and a half weeks and when 2500 people are infected.

17. If the derivative of the dose-response curve is smaller, the slope is not as steep. Since the slope is not as steep, the response increases less at the same dosage. Therefore, there is a wider range of dosages that are both safe and effective, and consequently the dosage given to the patient does not have to be as exact.

Solutions for Section 4.8

1. (a) See Figure 4.34.

(b) The surge function $y = ate^{bt}$ changes from increasing to decreasing at $t = \frac{1}{b}$. For this function $b = 0.2$ so the peak is at $\frac{1}{0.2} = 5$ hours. We can now substitute this into the formula to compute the peak concentration:

$$C = 12.4(5)e^{-0.2(5)} = 22.8085254 \text{ ng/ml} \approx 22.8 \text{ ng/ml}.$$

(c) Tracing along the graph of $C = 12.4te^{-0.2t}$, we see it crosses the line $C = 10$ at $t \approx 1$ hour and at $t \approx 14.4$ hours. Thus, the drug is effective for $1 \le t \le 14.4$ hours.

(d) The drug drops below $C = 4$ for $t > 20.8$ hours. Thus, it is safe to take the other drug after 20.8 hours.

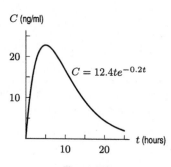

Figure 4.34

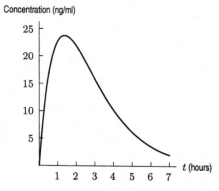

Figure 4.35

5. Figure 4.35 has its maximum at $t = 1.3$ hours, $C = 23.6$ ng/ml.

9. Food dramatically increases the value of the peak concentration but does not affect the time it takes to reach the peak concentration. The effect of food is stronger during the first 8 hours.

Solutions for Chapter 4 Review

1. See Figure 4.36.

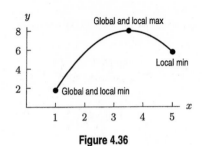

Figure 4.36

5. There are several possibilities. The price could have been increasing during the last few days of June, reaching a high point on July 1, then going back down during the first few days of July. In this case there was a local maximum in the price on July 1.

The price could have been decreasing during the last few days of June, reaching a low point on July 1, then going back up during the first few days of July. In this case there was a local minimum in the price on July 1.

It is also possible that there was neither a local maximum nor a local minimum in the price on July 1. This could have happened two ways. On the one hand, the price could have been rising in late June, then held steady with no change around July 1, after which the price increased some more. On the other hand, the price could have been falling in late June, then held steady with no change around July 1, after which the price fell some more. The key feature in these critical point scenarios is that there was no appreciable change in the price of the stock around July 1.

9. (a) Decreasing for $x < 0$, increasing for $0 < x < 4$, and decreasing for $x > 4$.
  (b) $f(0)$ is a local minimum, and $f(4)$ is a local maximum.

13. (a) The inflection point occurs at about week 14 where the graph changes from concave up to concave down.
  (b) The fetus increases its length faster at week 14 than at any other time during its gestation.

17. The derivative of $f(x) = x^5 + x + 7$ is $f'(x) = 5x^4 + 1$. Since $f'(x)$ is defined for all x and $f'(x) \neq 0$ for any value of x, there are no critical points for the function. Furthermore, $f'(x)$ is positive for all x, so the function is increasing over its entire domain, and hence can only cross the x-axis once. Since $f(x) \to +\infty$ as $x \to +\infty$ and $f(x) \to -\infty$ as $x \to -\infty$, the graph of f must cross the x-axis at least once, so we conclude that $f(x)$ has one real root.

21. (a) Quadratic polynomial (degree 2) with negative leading coefficient.
  (b) Exponential.
  (c) Logistic.
  (d) Logarithmic.
  (e) This is a quadratic polynomial (degree 2) and positive leading coefficient.
  (f) Exponential.
  (g) Surge
  (h) Periodic
  (i) This looks like a cubic polynomial (degree 3) and negative leading coefficient.

25. (a) $\pi(q)$ is maximized when $R(q) > C(q)$ and they are as far apart as possible. See Figure 4.37.
  (b) $\pi'(q_0) = R'(q_0) - C'(q_0) = 0$ implies that $C'(q_0) = R'(q_0) = p$.
      Graphically, the slopes of the two curves at q_0 are equal. This is plausible because if $C'(q_0)$ were greater than p or less than p, the maximum of $\pi(q)$ would be to the left or right of q_0, respectively. In economic terms, if the cost were rising more quickly than revenues, the profit would be maximized at a lower quantity (and if the cost were rising more slowly, at a higher quantity).
  (c) See Figure 4.38.

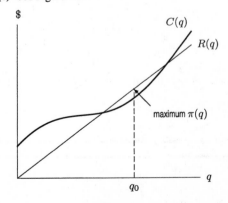

Figure 4.37

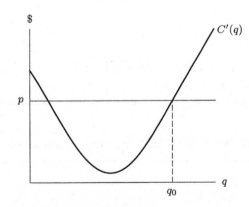

Figure 4.38

29. $C'(q)$ decreases with q because $C(q)$ is concave down. Therefore, $C'(2)$ is larger then $C'(3)$.

33. Since $C'(3)$ is the slope at $q = 3$, and $C(3)/3$ is the slope of the line from the origin to the point $q = 3$ on the curve $C(q)$, the value of $C(3)/3$ is larger.

37. Since $R = pq$, we have $dR/dp = p(dq/dp) + q$. We are assuming that

$$0 \le E = \left| \frac{p}{q} \cdot \frac{dq}{dp} \right| < 1$$

so, removing the absolute values

$$0 \ge -E = \frac{p}{q} \frac{dq}{dp} > -1$$

Multiplication by q gives

$$p \frac{dq}{dp} > -q$$

and hence

$$\frac{dR}{dp} = p \frac{dq}{dp} + q > 0$$

41. We rewrite $h(z)$ as $h(z) = z^{-1} + 4z^2$.
Differentiating gives

$$h'(z) = -z^{-2} + 8z,$$

so the critical points satisfy

$$-z^{-2} + 8z = 0$$
$$z^{-2} = 8z$$
$$8z^3 = 1$$
$$z^3 = \frac{1}{8}$$
$$z = \frac{1}{2}.$$

Since h' is negative for $0 < z < 1/2$ and h' is positive for $z > 1/2$, there is a local minimum at $z = 1/2$.

Since $h(z) \to \infty$ as $z \to 0^+$ and as $z \to \infty$, the local minimum at $z = 1/2$ is a global minimum; there is no global maximum. See Figure 4.39. Thus, the global minimum is $h(1/2) = 3$.

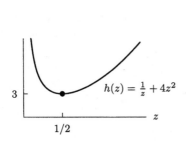

Figure 4.39

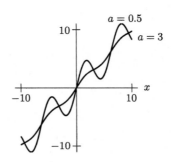

Figure 4.40

45. (a) See Figure 4.40.
  (b) The function $f(x) = x + a\sin x$ is increasing for all x if $f'(x) > 0$ for all x. We have $f'(x) = 1 + a\cos x$. Because $\cos x$ varies between -1 and 1, we have $1 + a\cos x > 0$ for all x if $-1 < a < 1$ but not otherwise. When $a = 1$, the function $f(x) = x + \sin x$ is increasing for all x, as is $f(x) = x - \sin x$, obtained when $a = -1$. Thus $f(x)$ is increasing for all x if $-1 \le a \le 1$.

49. (a) See Figure 4.41. The capacity appears to be 200 cars. The parking lot is full just before 8:30 am.
   (b) The rate of arrival between 5:00 and 5:30 is $(5 - 4)/0.5 = 2$ cars/hours. We assign this the time 5:15 (rather than either 5:00 or 5:30). The other values in Table 4.1 are calculated in a similar way. The data is plotted in Figure 4.42.

Table 4.1

Time (am)	5:15	5:45	6:15	6:45	7:15	7:45	8:15	8:45
Rate of arrival (cars/hour)	2	6	20	64	120	120	60	0

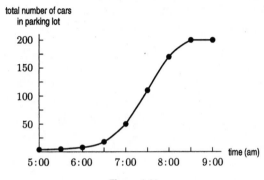

Figure 4.41

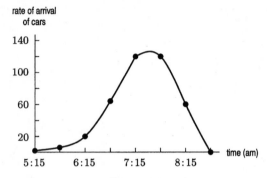

Figure 4.42

   (c) Rush hour occurs around the time when the rate of arrival of cars is maximum, namely, about 7:30 am.
   (d) The maximum of the rate of arrival occurs at the inflection point of the total number of cars in the lot.

53. (a) The vertical intercept is $W = Ae^{-e^{b-c\cdot 0}} = Ae^{-e^b}$. There is no horizontal intercept since the exponential function is always positive. There is a horizontal asymptote. As $t \to \infty$, we see that $e^{b-ct} = e^b/e^{ct} \to 0$, since t is positive. Therefore $W \to Ae^0 = A$, so there is a horizontal asymptote at $W = A$.
   (b) The derivative is

$$\frac{dW}{dt} = Ae^{-e^{b-ct}}(-e^{b-ct})(-c) = Ace^{-e^{b-ct}}e^{b-ct}.$$

Thus, dW/dt is always positive, so W is always increasing and has no critical points. The second derivative is

$$\frac{d^2W}{dt^2} = \frac{d}{dt}(Ace^{-e^{b-ct}})e^{b-ct} + Ace^{-e^{b-ct}}\frac{d}{dt}(e^{b-ct})$$

$$= Ac^2e^{-e^{b-ct}}e^{b-ct}e^{b-ct} + Ace^{-e^{b-ct}}(-c)e^{b-ct}$$

$$= Ac^2e^{-e^{b-ct}}e^{b-ct}(e^{b-ct} - 1).$$

Now e^{b-ct} decreases from $e^b > 1$ when $t = 0$ toward 0 as $t \to \infty$. The second derivative changes sign from positive to negative when $e^{b-ct} = 1$, i.e., when $b - ct = 0$, or $t = b/c$. Thus the curve has an inflection point at $t = b/c$, where $W = Ae^{-e^{b-(b/c)c}} = Ae^{-1}$.
   (c) See Figure 4.43.

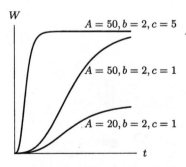

$A = 50, b = 2, c = 5$

$A = 50, b = 2, c = 1$

$A = 20, b = 2, c = 1$

Figure 4.43

   (d) The final size of the organism is given by the horizontal asymptote $W = A$. The curve is steepest at its inflection point, which occurs at $t = b/c$, $W = Ae^{-1}$. Since $e = 2.71828\ldots \approx 3$, the size the organism when it is growing fastest is about $A/3$, one third its final size. So yes, the Gompertz growth function is useful in modeling such growth.

CHAPTER FIVE

Solutions for Section 5.1

1. (a) Lower estimate $= (45)(2) + (16)(2) + (0)(2) = 122$ feet.
Upper estimate $= (88)(2) + (45)(2) + (16)(2) = 298$ feet.

(b)

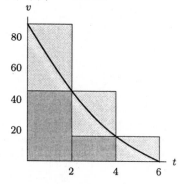

5. (a) Since car B starts at $t = 2$, the tick marks on the horizontal axis (which we assume are equally spaced) are 2 hours apart. Thus car B stops at $t = 6$ and travels for 4 hours.

Car A starts at $t = 0$ and stops at $t = 8$, so it travels for 8 hours.

(b) Car A's maximum velocity is approximately twice that of car B, that is 100 km/hr.

(c) The distance traveled is given by the area of under the velocity graph. Using the formula for the area of a triangle, the distances are given approximately by

$$\text{Car } A \text{ travels} = \frac{1}{2} \cdot \text{Base} \cdot \text{Height} = \frac{1}{2} \cdot 8 \cdot 100 = 400 \text{ km}$$

$$\text{Car } B \text{ travels} = \frac{1}{2} \cdot \text{Base} \cdot \text{Height} = \frac{1}{2} \cdot 4 \cdot 50 = 100 \text{ km}.$$

9. (a) Let's begin by graphing the data given in the table; see Figure 5.1. The total amount of pollution entering the lake during the 30-day period is equal to the shaded area. The shaded area is roughly 40% of the rectangle measuring 30 units by 35 units. Therefore, the shaded area measures about $(0.40)(30)(35) = 420$ units. Since the units are kilograms, we estimate that 420 kg of pollution have entered the lake.

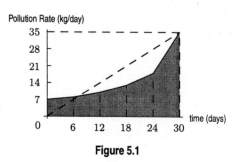

Figure 5.1

(b) Using left and right sums, we have

$$\text{Underestimate} = (7)(6) + (8)(6) + (10)(6) + (13)(6) + (18)(6) = 336 \text{ kg}.$$

$$\text{Overestimate} = (8)(6) + (10)(6) + (13)(6) + (18)(6) + (35)(6) = 504 \text{ kg}.$$

13. Using the data in Table 5.3 of Example 4, we construct Figure 5.2.

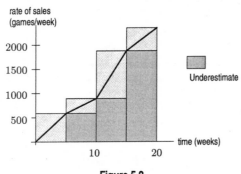

Figure 5.2

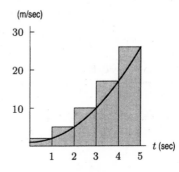

Figure 5.3

17. Sketch the graph of $v(t)$. See Figure 5.3. Adding up the areas using an overestimate with data every 1 second, we get $s \approx 2 + 5 + 10 + 17 + 26 = 60$ m. The actual distance traveled is less than 60 m.

Solutions for Section 5.2

1. Calculating both the LHS and RHS and averaging the two, we get

$$\frac{1}{2}(5(100 + 82 + 69 + 60 + 53) + 5(82 + 69 + 60 + 53 + 49)) = 1692.5$$

5.

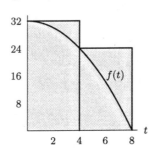

Figure 5.4: Left Sum, $\Delta t = 4$

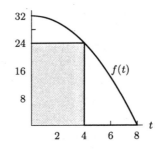

Figure 5.5: Right Sum, $\Delta t = 4$

(a) Left-hand sum $= 32 \cdot 4 + 24 \cdot 4 = 224$.
(b) Right-hand sum $= 24 \cdot 4 + 0 \cdot 4 = 96$.

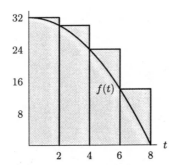

Figure 5.6: Left Sum, $\Delta t = 2$

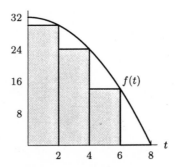

Figure 5.7: Right Sum, $\Delta t = 2$

(c) Left-hand sum $= 32 \cdot 2 + 30 \cdot 2 + 24 \cdot 2 + 14 \cdot 2 = 200$.
(d) Right-hand sum $= 30 \cdot 2 + 24 \cdot 2 + 14 \cdot 2 + 0 \cdot 2 = 136$.

9. $\int_0^3 f(x)\,dx$ is equal to the area shaded. We can use Riemann sum to estimate this area, or we can count grid squares. These are 3 whole grid squares and about 4 half-grid squares, for a total of 5 grid squares. Since each grid square represent 4 square units, our estimated area is $5(4) = 20$. We have $\int_0^3 f(x)\,dx \approx 20$. See Figure 5.8.

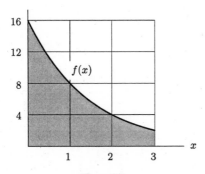

Figure 5.8

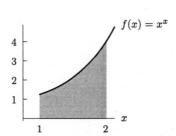

Figure 5.9

13. (a) See Figure 5.9. The shaded area appears to be approximately 2 units, and so $\int_1^2 x^x\,dx \approx 2$.

(b) $\int_1^2 x^x\,dx = 2.05045$

17. The graph given shows that f is positive for $0 \le t \le 1$. Since the graph is contained within a rectangle of height 100 and length 1, the answers -98.35 and 100.12 are both either too small or too large to represent $\int_0^1 f(t)dt$. Since the graph of f is above the horizontal line $y = 80$ for $0 \le t \le 0.95$, the best estimate is 93.47 and not 71.84.

21. $\int_{-1}^1 \frac{1}{e^t}\,dt = 2.350$

25. The integral $\int_1^3 \ln x\,dx \approx 1.30$

29. (a) With $n = 4$, we have $\Delta t = 4$. Then

$$t_0 = 0, t_1 = 4, t_2 = 8, t_3 = 12, t_4 = 16 \quad \text{and} \quad f(t_0) = 25, f(t_1) = 23, f(t_2) = 22, f(t_3) = 20, f(t_4) = 17$$

(b)

$$\text{Left sum} = (25)(4) + (23)(4) + (22)(4) + (20)(4) = 360$$
$$\text{Right sum} = (23)(4) + (22)(4) + (20)(4) + (17)(4) = 328.$$

(c) With $n = 2$, we have $\Delta t = 8$. Then

$$t_0 = 0, t_1 = 8, t_2 = 16 \quad \text{and} \quad f(t_0) = 25, f(t_1) = 22, f(t_2) = 17$$

(d)

$$\text{Left sum} = (25)(8) + (22)(8) = 376$$
$$\text{Right sum} = (22)(8) + (17)(8) = 312.$$

Solutions for Section 5.3

1. See Figure 5.10.

$$\text{Area} = \int_0^8 100(0.6)^t\,dt \approx 192.47$$

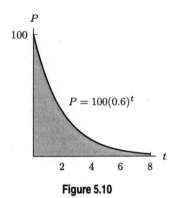

$$P = 100(0.6)^t$$

Figure 5.10

5. The entire graph of the function lies above the x-axis, so the integral is positive.

9. We know that

$$\int_{-3}^{5} f(x)dx = \text{Area above the axis} - \text{Area below the axis}.$$

The area above the axis is about 3 boxes. Since each box has area $(1)(5) = 5$, the area above the axis is about $(3)(5) = 15$. The area below the axis is about 11 boxes, giving an area of about $(11)(5) = 55$. We have

$$\int_{-3}^{5} f(x)dx \approx 15 - 55 = -40.$$

13. The area below the x-axis is bigger than the area above the x-axis, so the integral is negative. The area above the x-axis appears to be about half the area of the rectangle with area $20 \cdot 2 = 40$, so we estimate the area above the x-axis to be approximately 20. The area below the x-axis appears to be about half the area of the rectangle with area $15 \cdot 3 = 45$, so we estimate the area below the x-axis to be approximately 22.5. See Figure 5.11. The value of the integral is the area above the x-axis minus the area below the x-axis, so we estimate

$$\int_{0}^{5} f(x)dx \approx 20 - 22.5 = -2.5.$$

The correct match for this function is II.

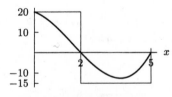

Figure 5.11

17. In Figure 5.12 the area A_1 is largest, A_2 is next, and A_3 is smallest. We have

$$\text{I} = \int_{a}^{b} f(x)\,dx = A_1, \quad \text{II} = \int_{a}^{c} f(x)\,dx = A_1 - A_2, \quad \text{III} = \int_{a}^{e} f(x)\,dx = A_1 - A_2 + A_3,$$

$$\text{IV} = \int_{b}^{e} f(x)\,dx = -A_2 + A_3, \quad \text{V} = \int_{b}^{c} f(x)\,dx = -A_2.$$

The relative sizes of A_1, A_2, and A_3 mean that I is positive and largest, III is next largest (since $-A_2 + A_3$ is negative, but less negative than $-A_2$), II is next largest, but still positive (since A_1 is larger than A_2). The integrals IV and V are both negative, but V is more negative. Thus

$$\text{V} < \text{IV} < 0 < \text{II} < \text{III} < \text{I}.$$

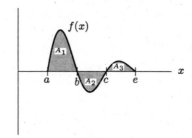

$f(x)$

A_1

a b A_2 c A_3 e x

Figure 5.12

21. **(a)** $\displaystyle\int_{-3}^{0} f(x)\,dx = -2.$

(b) $\displaystyle\int_{-3}^{4} f(x)\,dx = \int_{-3}^{0} f(x)\,dx + \int_{0}^{3} f(x)\,dx + \int_{3}^{4} f(x)\,dx = -2 + 2 - \frac{A}{2} = -\frac{A}{2}.$

25. A graph of $y = 6x^3 - 2$ shows that this function is nonnegative on the interval $x = 5$ to $x = 10$. Thus,

$$\text{Area} = \int_{5}^{10} (6x^3 - 2)\,dx = 14{,}052.5.$$

The integral was evaluated on a calculator.

29. The graph of $y = \cos x + 7$ is above $y = \ln(x - 3)$ for $5 \le x \le 7$. See Figure 5.13. Therefore

$$\text{Area} = \int_{5}^{7} \cos x + 7 - \ln(x - 3)\,dx = 13.457.$$

The integral was evaluated on a calculator.

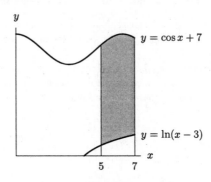

y

$y = \cos x + 7$

$y = \ln(x - 3)$

x

5   7

Figure 5.13

Solutions for Section 5.4

1. **(a)** The integral $\int_{0}^{30} f(t)\,dt$ represents the total emissions of nitrogen oxides, in millions of metric tons, during the period 1970 to 2000.

(b) We estimate the integral using left- and right-hand sums:

$$\text{Left sum} = (26.9)5 + (26.4)5 + (27.1)5 + (25.8)5 + (25.5)5 + (25.0)5 = 783.5.$$

$$\text{Right sum} = (26.4)5 + (27.1)5 + (25.8)5 + (25.5)5 + (25.0)5 + (22.6)5 = 762.0.$$

We average the left- and right-hand sums to find the best estimate of the integral:

$$\int_{0}^{30} f(t)\,dt \approx \frac{783.5 + 762.0}{2} = 772.8 \text{ million metric tons.}$$

Between 1970 and 2000, about 772.8 million metric tons of nitrogen oxides were emitted.

5. The integral $\int_{2000}^{2004} f(t)\, dt$ represents the change in the world's population between the years 2000 and 2004. It is measured in billions of people.

9. The total amount of antibodies produced is

$$\text{Total antibodies} = \int_0^4 r(t)dt \approx 1.417 \text{ thousand antibodies}$$

13. The change in the amount of water is the integral of rate of change, so we have

$$\text{Number of liters pumped out} = \int_0^{60} (5 - 5e^{-0.12t})dt = 258.4 \text{ liters.}$$

Since the tank contained 1000 liters of water initially, we see that

$$\text{Amount in tank after one hour} = 1000 - 258.4 = 741.6 \text{ liters.}$$

17. (a) In the beginning, both birth and death rates are small; this is consistent with a very small population. Both rates begin climbing, the birth rate faster than the death rate, which is consistent with a growing population. The birth rate is then high, but it begins to decrease as the population increases.

(b)

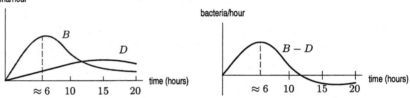

Figure 5.14: Difference between B and D is greatest at $t \approx 6$

The bacteria population is growing most quickly when $B - D$, the rate of change of population, is maximal; that happens when B is farthest above D, which is at a point where the slopes of both graphs are equal. That point is $t \approx 6$ hours.

(c) Total number born by time t is the area under the B graph from $t = 0$ up to time t. See Figure 5.15.

Total number alive at time t is the number born minus the number that have died, which is the area under the B graph minus the area under the D graph, up to time t. See Figure 5.16.

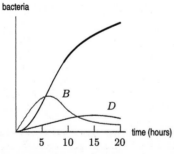

Figure 5.15: Number born by time t is $\int_0^t B(x)\, dx$

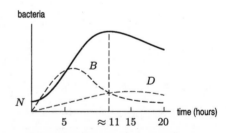

Figure 5.16: Number alive at time t is $\int_0^t (B(x) - D(x))\, dx$

From Figure 5.16, we see that the population is at a maximum when $B = D$, that is, after about 11 hours. This stands to reason, because $B - D$ is the rate of change of population, so population is maximized when $B - D = 0$, that is, when $B = D$.

21. Since W is in tons per week and t is in weeks since January 1, 2005, the integral $\int_0^{52} W\, dt$ gives the amount of waste, in tons, produced during the year 2005.

$$\text{Total waste during the year} = \int_0^{52} 3.75e^{-0.008t}\, dt = 159.5249 \text{ tons.}$$

Since waste removal costs \$15/ton, the cost of waste removal for the company is $159.5249 \cdot 15 = \$2392.87$.

25. The velocity is constant and negative, so the change in position is $-3 \cdot 5$ cm, that is 15 cm to the left.

29. From $t = 0$ to $t = 3$, you are moving away from home ($v > 0$); thereafter you move back toward home. So you are the farthest from home at $t = 3$. To find how far you are then, we can measure the area under the v curve as about 9 squares, or $9 \cdot 10$ km/hr $\cdot$ 1 hr $= 90$ km. To find how far away from home you are at $t = 5$, we measure the area from $t = 3$ to $t = 5$ as about 25 km, except that this distance is directed toward home, giving a total distance from home during the trip of $90 - 25 = 65$ km.

33.

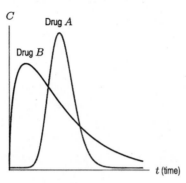

Figure 5.17

37. (a) The distance traveled is the integral of the velocity, so in T seconds you fall

$$\int_0^T 49(1 - 0.8187^t) \, dt.$$

(b) We want the number T for which

$$\int_0^T 49(1 - 0.8187^t) \, dt = 5000.$$

We can use a calculator or computer to experiment with different values for T, and we find $T \approx 107$ seconds.

Solutions for Section 5.5

1. The units for the integral $\int_{800}^{900} C'(q) dq$ are $\left(\frac{\text{dollars}}{\text{tons}}\right) \cdot (\text{tons}) = \text{dollars}$.

$\int_{800}^{900} C'(q) dq$ represents the cost of increasing production from 800 tons to 900 tons.

5. The area between 1970 and 1990 is about 15.3 grid squares, each of which has area $0.1(5) = 0.5$ million people. So

$$\text{Change in population} = \int_{1970}^{1990} P'(t) \, dt \approx 15.3(0.5) \approx 7.65 \text{ million people.}$$

The population of Tokyo increased by about 8 million people between 1970 and 1990.

9. (a)

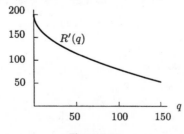

Figure 5.18

(b) By the Fundamental Theorem,

$$\int_0^{100} R'(q)dq = R(100) - R(0).$$

$R(0) = 0$ because no revenue is produced if no units are sold. Thus we get

$$R(100) = \int_0^{100} R'(q)dq \approx \$12{,}000.$$

(c) The marginal revenue in selling the 101st unit is given by $R'(100) = \$80/\text{unit}$. The total revenue in selling 101 units is:

$$R(100) + R'(100) = \$12{,}080.$$

13. We find the changes in $f(x)$ between any two values of x by counting the area between the curve of $f'(x)$ and the x-axis. Since $f'(x)$ is linear throughout, this is quite easy to do. From $x = 0$ to $x = 1$, we see that $f'(x)$ outlines a triangle of area $1/2$ below the x-axis (the base is 1 and the height is 1). By the Fundamental Theorem,

$$\int_0^1 f'(x)\,dx = f(1) - f(0),$$

so

$$f(0) + \int_0^1 f'(x)\,dx = f(1)$$

$$f(1) = 2 - \frac{1}{2} = \frac{3}{2}$$

Similarly, between $x = 1$ and $x = 3$ we can see that $f'(x)$ outlines a rectangle below the x-axis with area -1, so $f(2) = 3/2 - 1 = 1/2$. Continuing with this procedure (note that at $x = 4$, $f'(x)$ becomes positive), we get the table below.

x	0	1	2	3	4	5	6
$f(x)$	2	3/2	1/2	$-1/2$	-1	$-1/2$	1/2

Solutions for Chapter 5 Review

1. (a) Since the velocity is decreasing, for an upper estimate, we use a left sum. With $n = 5$, we have $\Delta t = 2$. Then

$$\text{Upper estimate} = (44)(2) + (42)(2) + (41)(2) + (40)(2) + (37)(2) = 408.$$

(b) For a lower estimate, we use a right sum, so

$$\text{Lower estimate} = (42)(2) + (41)(2) + (40)(2) + (37)(2) + (35)(2) = 390.$$

5. The units of measurement are meters per second (which are units of velocity).

9. $\int_0^{10} 2^{-x}\,dx = 1.44$

13. $\int_2^3 \frac{-1}{(r+1)^2}\,dr = -0.083$

17. Since $x^{1/2} \le x^{1/3}$ for $0 \le x \le 1$, we have

$$\text{Area} = \int_0^1 (x^{1/3} - x^{1/2})\,dx = 0.0833.$$

The integral was evaluated on a calculator.

21. To find the distance the car moved before stopping, we estimate the distance traveled for each two-second interval. Since speed decreases throughout, we know that the left-handed sum will be an overestimate to the distance traveled and the right-hand sum an underestimate. Applying the formulas for these sums with $\Delta t = 2$ gives:

$$\text{LEFT} = 2(100 + 80 + 50 + 25 + 10) = 530 \text{ ft.}$$
$$\text{RIGHT} = \ \ 2(80 + 50 + 25 + 10 + 0) \ \ = 330 \text{ ft.}$$

(a) The best estimate of the distance traveled will be the average of these two estimates, or

$$\text{Best estimate} = \frac{530 + 330}{2} = 430 \text{ ft.}$$

(b) All we can be sure of is that the distance traveled lies between the upper and lower estimates calculated above. In other words, all the black-box data tells us for sure is that the car traveled between 330 and 530 feet before stopping. So we can't be completely sure about whether it hit the skunk or not.

25. (a) The area under the curve is greater for species B for the first 5 years. Thus species B has a larger population after 5 years. After 10 years, the area under the graph for species B is still greater so species B has a greater population after 10 years as well.

(b) Unless something happens that we cannot predict now, species A will have a larger population after 20 years. It looks like species A will continue to quickly increase, while species B will add only a few new plants each year.

29. (a) Negative, since $f \leq 0$ everywhere on the interval $-5 \leq x \leq -4$.

(b) Positive, since $f \geq 0$ everywhere on the interval $-4 \leq x \leq 1$.

(c) Negative, since the graph of f has more area under the x-axis than above on the interval $1 \leq x \leq 3$.

(d) Positive, since the graph of f has more area above the x-axis than underneath on the interval $-5 \leq x \leq 3$.

33. (a) Since velocity is the rate of change of distance, we have

$$\text{Distance traveled} = \int_0^5 (10 + 8t - t^2) \, dt.$$

This distance is the shaded area in Figure 5.19.

(b) A graph of this velocity function is given in Figure 5.19. Finding the distance traveled is equivalent to finding the area under this curve between $t = 0$ and $t = 5$. We estimate that this area is about 100 since the average height appears to be about 20 and the width is 5.

(c) We use a calculator or computer to calculate the definite integral

$$\text{Distance traveled} = \int_0^5 (10 + 8t - t^2) \, dt \approx 108.33 \quad \text{meters.}$$

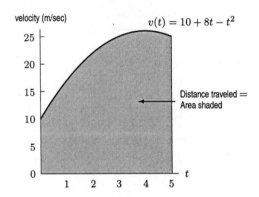

Figure 5.19: A velocity function

37. The area under the curve represents the number of cubic feet of storage times the number of days the storage was used. This area is given by

$$\text{Area under graph} = \text{Area of rectangle} + \text{Area of triangle}$$
$$= 30 \cdot 10{,}000 + \frac{1}{2} \cdot 30(30{,}000 - 10{,}000)$$
$$= 600{,}000.$$

Since the warehouse charges $5 for every 10 cubic feet of storage used for a day, the company will have to pay $(5)(60,000) = \$300,000$.

41. The graph of rate against time is the straight line shown in Figure 5.20. Since the shaded area is 270, we have

$$\frac{1}{2}(10 + 50) \cdot t = 270$$

$$t = \frac{270}{60} \cdot 2 = 9 \text{ years}$$

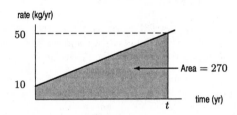

Figure 5.20

Solutions to Problems on the Second Fundamental Theorem of Calculus

1. By the Second Fundamental Theorem, $G'(x) = x^3$.

5. (a) If $F(b) = \int_0^b 2^x \, dx$ then $F(0) = \int_0^0 2^x \, dx = 0$ since we are calculating the area under the graph of $f(x) = 2^x$ on the interval $0 \le x \le 0$, or on no interval at all.

(b) Since $f(x) = 2^x$ is always positive, the value of F will increase as b increases. That is, as b grows larger and larger, the area under $f(x)$ on the interval from 0 to b will also grow larger.

(c) Using a calculator or a computer, we get

$$F(1) = \int_0^1 2^x \, dx \approx 1.4,$$

$$F(2) = \int_0^2 2^x \, dx \approx 4.3,$$

$$F(3) = \int_0^3 2^x \, dx \approx 10.1.$$

9. Note that $\int_a^b (g(x))^2 \, dx = \int_a^b (g(t))^2 \, dt$. Thus, we have

$$\int_a^b \left((f(x))^2 - (g(x))^2\right) \, dx = \int_a^b (f(x))^2 \, dx - \int_a^b (g(x))^2 \, dx = 12 - 3 = 9.$$

CHAPTER SIX

Solutions for Section 6.1

1. **(a)** Counting the squares yields an estimate of 25 squares, each with area $= 1$, so we conclude that

$$\int_0^5 f(x)\,dx \approx 25.$$

(b) The average height appears to be around 5.

(c) Using the formula, we get

$$\text{Average value} = \frac{\int_0^5 f(x)\,dx}{5-0} \approx \frac{25}{5} \approx 5,$$

which is consistent with (b).

5. **(a)** (i) Since the triangular region under the graphs of $f(x)$ has area $1/2$, we have

$$\text{Average}(f) = \frac{1}{2-0}\int_0^2 f(x)\,dx = \frac{1}{2}\cdot\frac{1}{2} = \frac{1}{4}.$$

(ii) Similarly,

$$\text{Average}(g) = \frac{1}{2-0}\int_0^2 g(x)\,dx = \frac{1}{2}\cdot\frac{1}{2} = \frac{1}{4}$$

(iii) Since $f(x)$ is nonzero only for $0 \le x < 1$ and $g(x)$ is nonzero only for $1 < x \le 2$, the product $f(x)g(x) = 0$ for all x. Thus

$$\text{Average}(f\cdot g) = \frac{1}{2-0}\int_0^2 f(x)g(x)\,dx = \frac{1}{2}\int_0^2 0\,dx = 0.$$

(b) Since the average values of $f(x)$ and $g(x)$ are nonzero, their product is nonzero. Thus the left side of the statement is nonzero. However, the average of the product $f(x)g(x)$ is zero. Thus, the right side of the statement is zero, so the statement is not true.

9. It appears that the area under a line at about $y = 2.5$ is approximately the same as the area under $f(x)$ on the interval $x = a$ to $x = b$, so we estimate that the average value is about 2.5. See Figure 6.1.

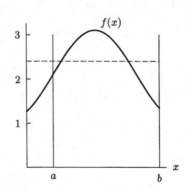

Figure 6.1

13. **(a)** Systolic, or maximum, blood pressure $= 120$ mm Hg.

(b) Diastolic, or minimum, blood pressure $= 80$ mm Hg.

(c) Average $= (120 + 80)/2 = 100$ mm Hg

(d) The pressure is at the diastolic level for a longer time than it is at or near the systolic level. Thus the average arterial pressure over the cardiac cycle is closer to the diastolic pressure than to the systolic pressure. The average pressure over the cycle is less than the average that was computed in part (c).

17. **(a)** Since $t = 0$ to $t = 31$ covers January:

$$\begin{array}{cc} \text{Average number of} \\ \text{daylight hours in January} \end{array} = \frac{1}{31} \int_0^{31} [12 + 2.4\sin(0.0172(t - 80))] \, dt.$$

Using left and right sums with $n = 100$ gives

$$\text{Average} \approx \frac{306}{31} \approx 9.9 \text{ hours.}$$

(b) Assuming it is not a leap year, the last day of May is $t = 151 (= 31 + 28 + 31 + 30 + 31)$ and the last day of June is $t = 181 (= 151 + 30)$. Again finding the integral numerically:

$$\begin{array}{cc} \text{Average number of} \\ \text{daylight hours in June} \end{array} = \frac{1}{30} \int_{151}^{181} [12 + 2.4\sin(0.0172(t - 80))] \, dt$$

$$\approx \frac{431}{30} \approx 14.4 \text{ hours.}$$

(c)

$$\text{Average for whole year} = \frac{1}{365} \int_0^{365} [12 + 2.4\sin(0.0172(t - 80))] \, dt$$

$$\approx \frac{4381}{365} \approx 12.0 \text{ hours.}$$

(d) The average over the whole year should be 12 hours, as computed in (c). Since Madrid is in the northern hemisphere, the average for a winter month, such as January, should be less than 12 hours (it is 9.9 hours) and the average for a summer month, such as June, should be more than 12 hours (it is 14.4 hours).

21. **(a)** $E(t) = 1.4e^{0.07t}$

(b)

$$\text{Average Yearly Consumption} = \frac{\text{Total Consumption for the Century}}{100 \text{ years}}$$

$$= \frac{1}{100} \int_0^{100} 1.4e^{0.07t} \, dt$$

$$\approx 219 \text{ million megawatt-hours.}$$

(c) We are looking for t such that $E(t) \approx 219$:

$$1.4e^{0.07t} \approx 219$$

$$e^{0.07t} = 156.4.$$

Taking natural logs,

$$0.07t = \ln 156.4$$

$$t \approx \frac{5.05}{0.07} \approx 72.18.$$

Thus, consumption was closest to the average during 1972.

(d) Between the years 1900 and 2000 the graph of $E(t)$ looks like

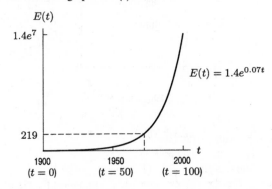

From the graph, we can see the t value such that $E(t) = 219$. It lies to the right of $t = 50$, and is thus in the second half of the century.

Solutions for Section 6.2

1. Looking at the graph we see that the supply and demand curves intersect at roughly the point $(345, 8)$. Thus the equilibrium price is \$8 per unit and the equilibrium quantity is 345 units. Figures 6.2 and 6.3 show the shaded areas corresponding to the consumer surplus and the producer surplus. Counting grid squares we see that the consumer surplus is roughly \$2000 while the producer surplus is roughly \$1400.

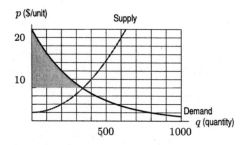

Figure 6.2: Consumer surplus

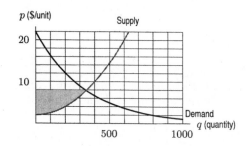

Figure 6.3: Producer surplus

5. When $q = 10$, the price is $p = 100 - 4 \cdot 10 = 60$. From Figure 6.20 on page 283 of the text, with $q^+ = 10$,

$$\text{Consumer surplus} = \int_0^{10} (100 - 4q)\, dq - 60 \cdot 10$$

Using a calculator or computer to evaluate the integral we get

$$\text{Consumer surplus} = 200.$$

9. **(a)** The supply and demand curves are shown in Figure 6.4. Tracing along the graphs, we find that they intersect approximately at the point $(322, 11.43)$. Thus, the equilibrium price is about \$11.43, and the equilibrium quantity is about 322 units.

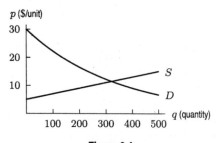

Figure 6.4

(b) The consumer surplus is shown in Figure 6.5. This is an area between two curves and we have

$$\text{Consumer surplus} = \int_0^{322} (30e^{-0.003q} - 11.43)\, dq = \$2513.52.$$

The producer surplus is shown in Figure 6.6. This is an area between two curves and we have

$$\text{Producer surplus} = \int_0^{322} (11.43 - (5 + 0.02q))\, dq = \$1033.62.$$

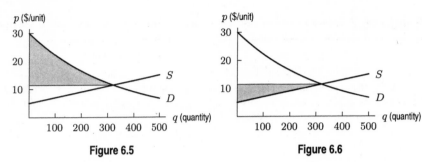

Figure 6.5   **Figure 6.6**

13. The supply curve, $S(q)$, represents the minimum price p per unit that the suppliers will be willing to supply some quantity q of the good for. See Figure 6.7. If the suppliers have q^* of the good and q^* is divided into subintervals of size Δq, then if the consumers could offer the suppliers for each Δq a price increase just sufficient to induce the suppliers to sell an additional Δq of the good, the consumers' total expenditure on q^* goods would be

$$p_1 \Delta q + p_2 \Delta q + \cdots = \sum p_i \Delta q.$$

As $\Delta q \to 0$ the Riemann sum becomes the integral $\int_0^{q^*} S(q)\, dq$. Thus $\int_0^{q^*} S(q)\, dq$ is the amount the consumers would pay if suppliers could be forced to sell at the lowest price they would be willing to accept.

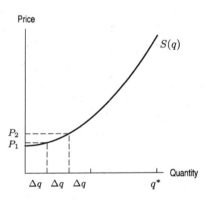

Figure 6.7

Solutions for Section 6.3

1. The present value is given by

$$\text{Present value} = \int_0^M S e^{-rt} dt,$$

with $S = 1000$, $r = 0.09$ and $M = 5$. Hence,

$$\text{Present value} = \int_0^5 1000 e^{-0.09t}\, dt = \$4026.35.$$

5. **(a)**   (i) If the interest rate is 3%, we have

$$\text{Present value} = \int_0^4 5000 e^{-0.03t}\, dt = \$18,846.59.$$

(ii) If the interest rate is 10%, we have

$$\text{Present value} = \int_0^4 5000e^{-0.10t}\, dt = \$16{,}484.00.$$

(b) At the end of the four-year period, if the interest rate is 3%,

$$\text{Value} = 18{,}846.59e^{0.03(4)} = \$21{,}249.47.$$

At 10%,

$$\text{Value} = 16{,}484.00e^{0.10(4)} = \$24{,}591.24.$$

9. We compute the present value of the company's earnings over the next 8 years:

$$\text{Present value of earnings} = \int_0^8 50{,}000e^{-0.07t}\, dt = \$306{,}279.24.$$

If you buy the rights to the earnings of the company for \$350,000, you expect that the earnings will be worth more than \$350,000. Since the present value of the earnings is less than this amount, you should not buy.

13. At any time t, the company receives income of $s(t) = 50e^{-t}$ thousands of dollars per year. Thus the present value is

$$
\begin{aligned}
\text{Present value} &= \int_0^2 s(t)e^{-0.06t}\, dt \\
&= \int_0^2 (50e^{-t})e^{-0.06t}\, dt \\
&= \$41{,}508.
\end{aligned}
$$

17. (a) Suppose the oil extracted over the time period $[0, M]$ is S. (See Figure 6.8.) Since $q(t)$ is the rate of oil extraction, we have:

$$S = \int_0^M q(t)\, dt = \int_0^M (a - bt)\, dt = \int_0^M (10 - 0.1t)\, dt.$$

To calculate the time at which the oil is exhausted, set $S = 100$ and try different values of M. We find $M = 10.6$ gives

$$\int_0^{10.6} (10 - 0.1t)\, dt = 100,$$

so the oil is exhausted in 10.6 years.

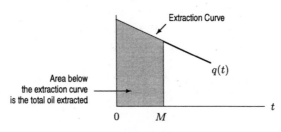

Figure 6.8

(b) Suppose p is the oil price, C is the extraction cost per barrel, and r is the interest rate. We have the present value of the profit as

$$
\begin{aligned}
\text{Present value of profit} &= \int_0^M (p - C)q(t)e^{-rt}\, dt \\
&= \int_0^{10.6} (20 - 10)(10 - 0.1t)e^{-0.1t}\, dt \\
&= 624.9 \text{ million dollars.}
\end{aligned}
$$

Solutions for Section 6.4

1. (a) Since the absolute growth rate for one year periods is constant, the function is linear, and the formula is
$$P = 100 + 10t.$$

    (b) Since the relative growth rate for one year periods is constant, the function is exponential, and the formula is
$$P = 100(1.10)^t.$$

    (c) See Figure 6.9.

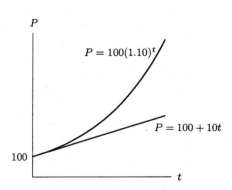

Figure 6.9: Two kinds of population growth

5. The change in $\ln P$ is the area under the curve, which is $\frac{1}{2} \cdot 10 \cdot (0.08) = 0.4$. So
$$\ln P(10) - \ln P(0) = \int_0^{10} \frac{P'(t)}{P(t)} \, dt = 0.4$$
$$\ln\left(\frac{P(10)}{P(0)}\right) = 0.4$$
$$\frac{P(10)}{P(0)} = e^{0.4} \approx 1.49.$$

    The population has increased by about 49% over the 10-year period.

9. Although the relative rate of change is decreasing, it is everywhere positive, so f is an increasing function for $0 \le t \le 10$.

13. (a) Let $P(t)$ be the population at time t in years. The population is increasing on the interval $0 \le t \le 10$. We have
$$\ln\left(\frac{P(10)}{P(0)}\right) = \int_0^{10} \frac{P'(t)}{P(t)} \, dt = \frac{1}{2} \cdot 10 \cdot 0.02 = 0.1.$$

    Therefore,
$$\frac{P(10)}{P(0)} = e^{0.1} = 1.105, \quad \text{so} \quad P(10) = 1.105 \cdot P(0).$$

    The population grew by about 10.5% during this time.

    (b) The population is decreasing on the interval $10 \le t \le 15$. We have
$$\ln\left(\frac{P(15)}{P(10)}\right) = \int_{10}^{15} \frac{P'(t)}{P(t)} \, dt = -\frac{1}{2} \cdot 5 \cdot 0.01 = -0.025.$$

    Therefore,
$$\frac{P(15)}{P(10)} = e^{-0.025} = 0.975, \quad \text{so} \quad P(15) = 0.975 \cdot P(10).$$

    The population decreased by about 2.5% during this time.

(c) On the entire interval, we have

$$\ln\left(\frac{P(15)}{P(0)}\right) = \int_0^{15} \frac{P'(t)}{P(t)} dt = 0.1 - 0.025 = 0.075.$$

Therefore,

$$\frac{P(15)}{P(0)} = e^{0.075} = 1.078, \quad \text{so} \quad P(15) = 1.078 \cdot P(0).$$

The population grew about 7.8% during this 15-year period.

17. (a) We compute the left- and right-hand sums:

$$\text{Left sum} = -0.03 + 0.03 - 0.06\ldots + 0.03 = -0.38$$

and

$$\text{Right sum} = 0.03 - 0.06\ldots + 0.03 + 0.02 = -0.33,$$

and average these to get our best estimate

$$\text{Average} = \frac{-0.38 + (-0.33)}{2} = -0.355.$$

We estimate that

$$\int_{1990}^{2002} \frac{P'(t)}{P(t)} dt \approx -0.36.$$

(b) The integral found in part (a) is the change in $\ln P(t)$, so we have

$$\ln P(2002) - \ln P(1990) = \int_{1990}^{2002} \frac{P'(t)}{P(t)} dt = -0.36$$

$$\ln\left(\frac{P(2002)}{P(1990)}\right) = -0.36$$

$$\frac{P(2002)}{P(1990)} = e^{-0.36} = 0.70.$$

Burglaries went down by a factor of 0.70 during this time, which means that the number of burglaries in 2002 was 0.70 times the number of burglaries in 1990. This is a decrease of about 30% during this period.

Solutions for Chapter 6 Review

1. By counting grid squares, we find $\int_1^6 f(x)\, dx = 8.5$, so the average value of f is $\frac{8.5}{6-1} = \frac{8.5}{5} = 1.7$.

5. (a) $Q(10) = 4(0.96)^{10} \approx 2.7.$ $Q(20) = 4(0.96)^{20} \approx 1.8$
   (b)
$$\frac{Q(10) + Q(20)}{2} \approx 2.21$$

   (c) The average value of Q over the interval is about 2.18.
   (d) Because the graph of Q is concave up between $t = 10$ and $t = 20$, the area under the curve is less than what is obtained by connecting the endpoints with a straight line.

9. We'll show that in terms of the average value of f,

$$\text{I} > \text{II} = \text{IV} > \text{III}$$

Using the definition of average value and the fact that f is even, we have

$$\frac{\text{Average value}}{\text{of } f \text{ on II}} = \frac{\int_0^2 f(x)dx}{2} = \frac{\frac{1}{2}\int_{-2}^2 f(x)dx}{2}$$

$$= \frac{\int_{-2}^2 f(x)dx}{4}$$

$$= \text{Average value of } f \text{ on IV}.$$

Since f is decreasing on $[0,5]$, the average value of f on the interval $[0, c]$, where $0 \le c \le 5$, is decreasing as a function of c. The larger the interval the more low values of f are included. Hence

$$\text{Average value of } f \text{ on } [0, 1] > \text{Average value of } f \text{ on } [0, 2] > \text{Average value of } f \text{ on } [0, 5]$$

13. (a) The quantity demanded at a price of $50 is calculated by substituting $p = 50$ into the demand equation $p = 100e^{-0.008q}$. Solving $50 = 100e^{-0.008q}$ for q gives $q \approx 86.6$. In other words, at a price of $50, consumer demand is about 87 units. The quantity supplied at a price of $50 is calculated by substituting by $p = 50$ into the supply equation $p = 4\sqrt{q} + 10$. Solving $50 = 4\sqrt{q} + 10$ for q gives $q = 100$. So at a price of $50, producers supply about 100 units. At a price of $50, the supply is larger than the demand, so some goods remain unsold. We can expect prices to be pushed down.

(b) The supply and demand curves are shown in Figure 6.10. The equilibrium price is about $p^* = \$48$ and the equilibrium quantity is about $q^* = 91$ units. The market will push prices downward from $50 toward the equilibrium price of $48. This agrees with the conclusion to part (a) that prices will drop.

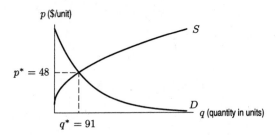

Figure 6.10: Demand and supply curves for a product

(c) See Figure 6.11. We have

$$\text{Consumer surplus} = \text{Area between demand curve and horizontal line } p = p^*.$$

The demand curve has equation $p = 100e^{0.008q}$, so

$$\text{Consumer surplus} = \int_0^{91} 100e^{-0.008q}\, dq - p^* q^* = 6464 - 48 \cdot 91 = 2096.$$

Consumers gain $2096 by buying goods at the equilibrium price instead of the price they would have been willing to pay.

For producer surplus, see Figure 6.12. We have

$$\text{Producer surplus} = \text{Area between supply curve and horizontal line } p = p^*$$

The supply curve has equation $p = 4\sqrt{q} + 10$, so

$$\text{Producer surplus} = p^* q^* - \int_0^{91} (4\sqrt{q} + 10)\, dq = 48 \cdot 91 - 3225 = 1143.$$

Producers gain $1143 by supplying goods at the equilibrium price instead of the price at which they would have been willing to provide the goods.

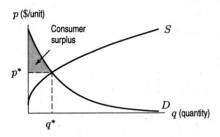

Figure 6.11: Consumer surplus

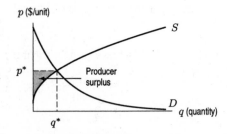

Figure 6.12: Producer surplus

17. January 1, 1996 through January 1, 2003 is a seven-year time period, and $t = 0$ corresponds to January 1, 1996, so the value on January 1, 1996 of the sales over this seven-year time period is

$$\text{Value on Jan. 1, 1996} = \int_0^7 1431 e^{0.134t} e^{-0.075t} \, dt$$

$$= \int_0^7 1431 e^{0.059t} \, dt$$

$$= 12{,}402.28 \quad \text{million dollars.}$$

The value, on January 1, 1996, of Harley-Davidson sales over the time period from January 1, 1996 through January 1, 2003 is about 12,402 million dollars.

21. (a) The town is growing by 50 people per year. The population in year 1 is thus $1000 + 50 = 1050$. Continuing to add 50 people per year to the population of the town produces the data in Table 6.1.

Table 6.1

Year	0	1	2	3	4	...	10
Population	1000	1050	1100	1150	1200	...	1500

(b) The town is growing at 5% per year. The population in year 1 is thus $1000 + 1000(0.05) = 1000(1.05) = 1050$. Continuing to multiply the town's population by 1.05 each year yields the data in Table 6.2.

Table 6.2

Year	0	1	2	3	4	...	10
Population	1000	1050	1103	1158	1216	...	1629

CHAPTER SEVEN

Solutions for Section 7.1

1. $5x$

5. $\dfrac{x^5}{5}$.

9. We break the antiderivative into two terms. Since y^3 is an antiderivative of $3y^2$ and $-y^4/4$ is an antiderivative of $-y^3$, an antiderivative of $3y^2 - y^3$ is

$$y^3 - \frac{y^4}{4}.$$

13. $t^3 + \dfrac{7t^2}{2} + t$.

17. $\dfrac{t^4}{4} - \dfrac{t^3}{6} - \dfrac{t^2}{2}$

21. $F(x) = \dfrac{x^7}{7} - \dfrac{1}{7}\left(\dfrac{x^{-5}}{-5}\right) + C = \dfrac{x^7}{7} + \dfrac{1}{35}x^{-5} + C$

25. $G(t) = 5t + \sin t + C$

29. $f(x) = x^2$, so $F(x) = \dfrac{x^3}{3} + C$. $F(0) = 0$ implies that $\dfrac{0^3}{3} + C = 0$, so $C = 0$. Thus $F(x) = \dfrac{x^3}{3}$ is the only possibility.

33. $\dfrac{3x^2}{2} + C$

37. $\dfrac{x^4}{4} - \dfrac{x^2}{2} + C$.

41. $\dfrac{q^3}{3} + \dfrac{5q^2}{2} + 2q + C$

45. $-\dfrac{5}{t} - \dfrac{3}{t^2} + C$

49. $\dfrac{-1}{0.05}e^{-0.05t} + C = -20e^{-0.05t} + C$.

53. $-\cos t + C$

57. $5\sin x + 3\cos x + C$

61. $10x - 4\cos(2x) + C$

65. An antiderivative is $F(x) = 2e^{3x} + C$. Since $F(0) = 5$, we have $5 = 2e^0 + C = 2 + C$, so $C = 3$. The answer is $F(x) = 2e^{3x} + 3$.

Solutions for Section 7.2

1. We use the substitution $w = x^2 + 1$, $dw = 2x\,dx$.

$$\int 2x(x^2 + 1)^5 dx = \int w^5 dw = \frac{w^6}{6} + C = \frac{1}{6}(x^2 + 1)^6 + C.$$

Check: $\dfrac{d}{dx}\left(\dfrac{1}{6}(x^2 + 1)^6 + C\right) = 2x(x^2 + 1)^5$.

5. We use the substitution $w = x^2 + 1$, $dw = 2x\,dx$.

$$\int \frac{2x}{\sqrt{x^2 + 1}}\,dx = \int w^{-1/2}\,dw = 2w^{1/2} + C = 2\sqrt{x^2 + 1} + C.$$

Check: $\dfrac{d}{dx}(2\sqrt{x^2 + 1} + C) = \dfrac{2x}{\sqrt{x^2 + 1}}.$

9. We use the substitution $w = t^3 - 3$, $dw = 3t^2\,dt$.

$$\int t^2(t^3 - 3)^{10}\,dt = \frac{1}{3}\int (t^3 - 3)^{10}(3t^2\,dt) = \int w^{10}\left(\frac{1}{3}\,dw\right)$$

$$= \frac{1}{3}\frac{w^{11}}{11} + C = \frac{1}{33}(t^3 - 3)^{11} + C.$$

Check: $\dfrac{d}{dt}[\dfrac{1}{33}(t^3 - 3)^{11} + C] = \dfrac{1}{3}(t^3 - 3)^{10}(3t^2) = t^2(t^3 - 3)^{10}.$

13. We use the substitution $w = 4 - x$, $dw = -dx$.

$$\int \frac{1}{\sqrt{4 - x}}\,dx = -\int \frac{1}{\sqrt{w}}\,dw = -2\sqrt{w} + C = -2\sqrt{4 - x} + C.$$

Check: $\dfrac{d}{dx}(-2\sqrt{4 - x} + C) = -2 \cdot \dfrac{1}{2} \cdot \dfrac{1}{\sqrt{4 - x}} \cdot -1 = \dfrac{1}{\sqrt{4 - x}}.$

17. In this case, it seems easier not to substitute.

$$\int (x^2 + 3)^2\,dx = \int (x^4 + 6x^2 + 9)\,dx = \frac{x^5}{5} + 2x^3 + 9x + C.$$

Check: $\dfrac{d}{dx}\left[\dfrac{x^5}{5} + 2x^3 + 9x + C\right] = x^4 + 6x^2 + 9 = (x^2 + 3)^2.$

21. We use the substitution $w = \cos 3t$, $dw = -3\sin 3t\,dt$.

$$\int \sqrt{\cos 3t}\,\sin 3t\,dt = -\frac{1}{3}\int \sqrt{w}\,dw$$

$$= -\frac{1}{3} \cdot \frac{2}{3}w^{\frac{3}{2}} + C = -\frac{2}{9}(\cos 3t)^{\frac{3}{2}} + C.$$

Check:

$$\frac{d}{dt}\left[-\frac{2}{9}(\cos 3t)^{\frac{3}{2}} + C\right] = -\frac{2}{9} \cdot \frac{3}{2}(\cos 3t)^{\frac{1}{2}} \cdot (-\sin 3t) \cdot 3$$

$$= \sqrt{\cos 3t}\,\sin 3t.$$

25. We use the substitution $w = \sin 5\theta$, $dw = 5\cos 5\theta\,d\theta$.

$$\int \sin^6 5\theta \cos 5\theta\,d\theta = \frac{1}{5}\int w^6\,dw = \frac{1}{5}(\frac{w^7}{7}) + C = \frac{1}{35}\sin^7 5\theta + C.$$

Check: $\dfrac{d}{d\theta}(\dfrac{1}{35}\sin^7 5\theta + C) = \dfrac{1}{35}[7\sin^6 5\theta](5\cos 5\theta) = \sin^6 5\theta \cos 5\theta.$

Note that we could also use Problem 23 to solve this problem, substituting $w = 5\theta$ and $dw = 5\,d\theta$ to get:

$$\int \sin^6 5\theta \cos 5\theta\,d\theta = \frac{1}{5}\int \sin^6 w \cos w\,dw$$

$$= \frac{1}{5}(\frac{\sin^7 w}{7}) + C = \frac{1}{35}\sin^7 5\theta + C.$$

29. We use the substitution $w = 3x^2$, $dw = 6x\,dx$.

$$\int xe^{3x^2}\,dx = \frac{1}{6}\int e^w\,dw = \frac{1}{6}e^w + C = \frac{1}{6}e^{3x^2} + C.$$

Check: $\dfrac{d}{dx}\left(\dfrac{1}{6}e^{3x^2} + C\right) = \dfrac{1}{6}e^{3x^2} \cdot 6x = xe^{3x^2}.$

33. We use the substitution $w = e^t + t$, $dw = (e^t + 1)\,dt$.

$$\int \frac{e^t + 1}{e^t + t}\,dt = \int \frac{1}{w}\,dw = \ln|w| + C = \ln|e^t + t| + C.$$

Check: $\dfrac{d}{dt}(\ln|e^t + t| + C) = \dfrac{e^t + 1}{e^t + t}.$

37. We use the substitution $w = x + e^x$, $dw = (1 + e^x)\,dx$.

$$\int \frac{1 + e^x}{\sqrt{x + e^x}}\,dx = \int \frac{dw}{\sqrt{w}} = 2\sqrt{w} + C = 2\sqrt{x + e^x} + C.$$

Check: $\dfrac{d}{dx}(2\sqrt{x + e^x} + C) = 2 \cdot \dfrac{1}{2}(x + e^x)^{-\frac{1}{2}} \cdot (1 + e^x) = \dfrac{1 + e^x}{\sqrt{x + e^x}}.$

41. (a) This integral can be evaluated using integration by substitution. We use $w = x^2$, $dw = 2x\,dx$.

$$\int x\sin x^2\,dx = \frac{1}{2}\int \sin(w)\,dw = -\frac{1}{2}\cos(w) + C = -\frac{1}{2}\cos(x^2) + C.$$

(b) This integral cannot be evaluated using a simple integration by substitution.

(c) This integral cannot be evaluated using a simple integration by substitution.

(d) This integral can be evaluated using integration by substitution. We use $w = 1 + x^2$, $dw = 2x\,dx$.

$$\int \frac{x}{(1 + x^2)^2}\,dx = \frac{1}{2}\int \frac{1}{w^2}\,dw = \frac{1}{2}\left(\frac{-1}{w}\right) + C = \frac{-1}{2(1 + x^2)} + C.$$

(e) This integral cannot be evaluated using a simple integration by substitution.

(f) This integral can be evaluated using integration by substitution. We use $w = 2 + \cos x$, $dw = -\sin x\,dx$.

$$\int \frac{\sin x}{2 + \cos x}\,dx = -\int \frac{1}{w}\,dw = -\ln|w| + C = -\ln|2 + \cos x| + C.$$

Solutions for Section 7.3

1. Since $F'(x) = 5$, we use $F(x) = 5x$. By the Fundamental Theorem, we have

$$\int_1^3 5\,dx = 5x\bigg|_1^3 = 5(3) - 5(1) = 15 - 5 = 10.$$

5. Since $F'(x) = \dfrac{1}{x^2} = x^{-2}$, we use $F(x) = \dfrac{x^{-1}}{-1} = -\dfrac{1}{x}$. By the Fundamental Theorem, we have

$$\int_1^2 \frac{1}{x^2}\,dx = \left(-\frac{1}{x}\right)\bigg|_1^2 = -\frac{1}{2} - \left(-\frac{1}{1}\right) = -\frac{1}{2} + 1 = \frac{1}{2}.$$

9. If $F'(x) = 6x^2$, then $F(x) = 2x^3$. By the Fundamental Theorem, we have

$$\int_1^3 6x^2\,dx = 2x^3\bigg|_1^3 = 2(27) - 2(1) = 54 - 2 = 52.$$

13. If $f(x) = 1/x$, then $F(x) = \ln|x|$ (since $\dfrac{d}{dx} \ln|x| = \dfrac{1}{x}$). By the Fundamental Theorem, we have

$$\int_1^2 \frac{1}{x}\, dx = \ln|x|\Big|_1^2 = \ln 2 - \ln 1 = \ln 2.$$

17. $\displaystyle\int_0^1 \sin\theta\, d\theta = -\cos\theta\Big|_0^1 = 1 - \cos 1 \approx 0.460.$

21. (a) We substitute $w = 1 + x^2$, $dw = 2x\, dx$.

$$\int_{x=0}^{x=1} \frac{x}{1+x^2}\, dx = \frac{1}{2}\int_{w=1}^{w=2} \frac{1}{w}\, dw = \frac{1}{2}\ln|w|\Big|_1^2 = \frac{1}{2}\ln 2.$$

(b) We substitute $w = \cos x$, $dw = -\sin x\, dx$.

$$\int_{x=0}^{x=\frac{\pi}{4}} \frac{\sin x}{\cos x}\, dx = -\int_{w=1}^{w=\sqrt{2}/2} \frac{1}{w}\, dw$$

$$= -\ln|w|\Big|_1^{\sqrt{2}/2} = -\ln\frac{\sqrt{2}}{2} = \frac{1}{2}\ln 2.$$

25. We substitute $w = t + 1$, so $dw = dt$.

$$\int \frac{1}{\sqrt{t+1}}\, dt = \int \frac{1}{\sqrt{w}}\, dw = \int w^{-1/2}\, dw = 2w^{1/2} + C = 2\sqrt{t+1} + C.$$

Using the Fundamental Theorem, we have

$$\int_0^3 \frac{1}{\sqrt{t+1}}\, dt = 2\sqrt{t+1}\Big|_0^3 = 2\sqrt{4} - 2\sqrt{1} = 4 - 2 = 2.$$

29. One antiderivative of $f(x) = e^{0.5x}$ is $F(x) = 2e^{0.5x}$. Thus, the definite integral of $f(x)$ on the interval $0 \le x \le 3$ is

$$\int_0^3 e^{0.5x}\, dx = F(3) - F(0) = 2e^{0.5x}\Big|_0^3.$$

The average value of a function on a given interval is the definite integral over that interval divided by the length of the interval:

$$\text{Average value} = \left(\frac{1}{3-0}\right)\cdot\left(\int_0^3 e^{0.5x}\, dx\right) = \frac{1}{3}\left(2e^{0.5x}\Big|_0^3\right) = \frac{1}{3}(2e^{1.5} - 2e^0) \approx 2.32.$$

From the graph of $y = e^{0.5x}$ in Figure 7.1 we see that an average value of 2.32 on the interval $0 \le x \le 3$ does make sense.

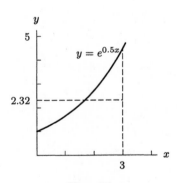

Figure 7.1

33. We have

$$\text{Area} = \int_0^b x^2 \, dx = \frac{x^3}{3}\Big|_0^b = \frac{b^3}{3}.$$

We find the value of b making the area equal to 100:

$$100 = \frac{b^3}{3}$$
$$300 = b^3$$
$$b = (300)^{1/3} = 6.694.$$

37. (a) We sketch $f(x) = xe^{-x}$; see Figure 7.2. The shaded area to the right of the y-axis represents the integral $\int_0^\infty xe^{-x} \, dx$.

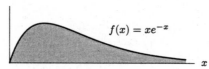

$f(x) = xe^{-x}$

Figure 7.2

(b) Using a calculator or computer, we obtain

$$\int_0^5 xe^{-x} \, dx = 0.9596 \quad \int_0^{10} xe^{-x} \, dx = 0.9995 \quad \int_0^{20} xe^{-x} \, dx = 0.99999996.$$

(c) The answers to part (b) suggest that the integral converges to 1.

41. (a) Evaluating the integrals with a calculator gives

$$\int_0^{10} xe^{-x/10} \, dx = 26.42$$

$$\int_0^{50} xe^{-x/10} \, dx = 95.96$$

$$\int_0^{100} xe^{-x/10} \, dx = 99.95$$

$$\int_0^{200} xe^{-x/10} \, dx = 100.00$$

(b) The results of part (a) suggest that

$$\int_0^\infty xe^{-x/10} \, dx \approx 100$$

45. Since $y = x^3(1 - x)$ is positive for $0 \le x \le 1$ and $y = 0$, when $x = 0, 1$, the area is given by

$$\text{Area} = \int_0^1 x^3(1 - x) \, dx = \int_0^1 (x^3 - x^4) \, dx = \frac{x^4}{4} - \frac{x^5}{5}\Big|_0^1 = \frac{1}{20}.$$

Solutions for Section 7.4

1. Since dP/dt is negative for $t < 3$ and positive for $t > 3$, we know that P is decreasing for $t < 3$ and increasing for $t > 3$. Between each two integer values, the magnitude of the change is equal to the area between the graph dP/dt and the t-axis. For example, between $t = 0$ and $t = 1$, we see that the change in P is -1. Since $P = 2$ at $t = 0$, we must have $P = 1$ at $t = 1$. The other values are found similarly, and are shown in Table 7.1.

Table 7.1

t	1	2	3	4	5
P	1	0	$-1/2$	0	1

5. See Figure 7.3.

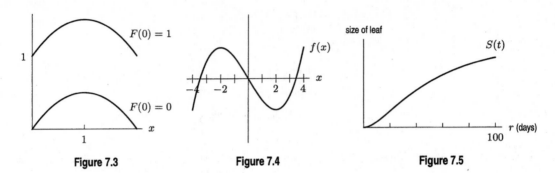

Figure 7.3   Figure 7.4   Figure 7.5

9. (a) The function $f(x)$ is increasing when $f'(x)$ is positive, so $f(x)$ is increasing for $x < -2$ or $x > 2$.
The function $f(x)$ is decreasing when $f'(x)$ is negative, so $f(x)$ is decreasing for $-2 < x < 2$.
Since $f(x)$ is increasing to the left of $x = -2$, decreasing between $x = -2$ and $x = 2$, and increasing to the right of $x = 2$, the function $f(x)$ has a local maximum at $x = -2$ and a local minimum at $x = 2$.

(b) See Figure 7.4.

13. Since the rate at which the leaf grows is proportional to the rate of photosynthesis, the slope of the size graph is proportional to the given graph. Thus, if $S(t)$ is the size of the leaf and $p(t)$ is the rate of photosynthesis

$$S'(t) = kp(t) \qquad \text{for some positive } k.$$

We plot the antiderivative of $p(t)$ to get the graph of $S(t)$ in Figure 7.5. (Since no scale is given on the vertical axis, we can imagine $k = 1$.) The size of the leaf may be represented by its area, or perhaps by its weight.

17. The critical points are at $(0, 5)$, $(2, 21)$, $(4, 13)$, and $(5, 15)$. A graph is given in Figure 7.6.

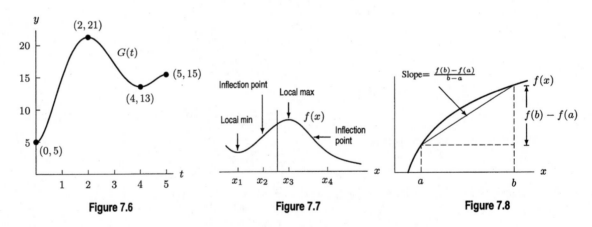

Figure 7.6   Figure 7.7   Figure 7.8

21. See Figure 7.7.

25. See Figure 7.8.

Solutions for Chapter 7 Review

1. $10x + 8(\dfrac{x^4}{4}) = 10x + 2x^4$.

5. $P(r) = \pi r^2 + C$

9. Antiderivative $G(x) = \dfrac{x^4}{4} + x^3 + \dfrac{3x^2}{2} + x + C = \dfrac{(x+1)^4}{4} + C$

13. $\displaystyle\int (x+1)^2 \, dx = \frac{(x+1)^3}{3} + C.$

Another way to work the problem is to expand $(x+1)^2$ to $x^2 + 2x + 1$ as follows:

$$\int (x+1)^2 \, dx = \int (x^2 + 2x + 1) \, dx = \frac{x^3}{3} + x^2 + x + C.$$

These two answers are the same, since $\dfrac{(x+1)^3}{3} = \dfrac{x^3 + 3x^2 + 3x + 1}{3} = \dfrac{x^3}{3} + x^2 + x + \dfrac{1}{3}$, which is $\dfrac{x^3}{3} + x^2 + x$, plus a constant.

17. $4t^2 + 3t + C.$

21. $2 \ln|x| - \pi \cos x + C$

25. If $F'(t) = \cos t$, we can take $F(t) = \sin t$, so

$$\int_{-1}^{1} \cos t \, dt = \sin t \Big|_{-1}^{1} = \sin 1 - \sin(-1).$$

Since $\sin(-1) = -\sin 1$, we can simplify the answer and write

$$\int_{-1}^{1} \cos t \, dt = 2 \sin 1$$

29. $f(x) = 2x$, so $F(x) = x^2 + C$. $F(0) = 0$ implies that $0^2 + C = 0$, so $C = 0$. Thus $F(x) = x^2$ is the only possibility.

33. We use the substitution $w = q^2 + 1$, $dw = 2q\,dq$.

$$\int 2q e^{q^2+1} dq = \int e^w dw = e^w + C = e^{q^2+1} + C.$$

37. We use the substitution $w = 5x - 7$, $dw = 5dx$.

$$\int (5x-7)^{10} dx = \frac{1}{5} \int w^{10} dw = \frac{1}{5} \frac{w^{11}}{11} + C = \frac{1}{55}(5x-7)^{11} + C.$$

41. We use the substitution $w = x^3$, $dw = 3x^2 dx$.

$$12 \int x^2 \cos(x^3) dx = \frac{12}{3} \int \cos(w) dw = 4\sin(w) + C = 4\sin(x^3) + C.$$

45. (a) Critical points of $F(x)$ are the zeros of f: $x = 1$ and $x = 3$.
   (b) $F(x)$ has a local minimum at $x = 1$ and a local maximum at $x = 3$.
   (c) See Figure 7.9.

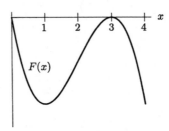

Figure 7.9

Notice that the graph could also be above or below the x-axis at $x = 3$.

49. (a) The graph of $y = e^{-x^2}$ is in Figure 7.10. The integral $\int_{-\infty}^{\infty} e^{-x^2} \, dx$ represents the entire area under the curve, which is shaded.

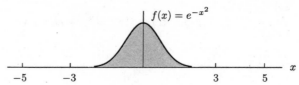

Figure 7.10

(b) Using a calculator or computer, we see that

$$\int_{-1}^{1} e^{-x^2} \, dx = 1.494, \quad \int_{-2}^{2} e^{-x^2} \, dx = 1.764, \quad \int_{-3}^{3} e^{-x^2} \, dx = 1.772, \quad \int_{-5}^{5} e^{-x^2} \, dx = 1.772$$

(c) From part (b), we see that as we extend the limits of integration, the area appears to get closer and closer to about 1.772. We estimate that

$$\int_{-\infty}^{\infty} e^{-x^2} \, dx = 1.772$$

Solutions to Practice Problems on Integration

1. $\displaystyle\int (t^3 + 6t^2) \, dt = \frac{t^4}{4} + 6 \cdot \frac{t^3}{3} + C = \frac{t^4}{4} + 2t^3 + C$

5. $\displaystyle\int 3w^{1/2} \, dw = 3 \cdot \frac{w^{3/2}}{3/2} + C = 2w^{3/2} + C$

9. $\displaystyle\int (w^4 - 12w^3 + 6w^2 - 10) \, dw = \frac{w^5}{5} - 12 \cdot \frac{w^4}{4} + 6 \cdot \frac{w^3}{3} - 10 \cdot w + C$

$$= \frac{w^5}{5} - 3w^4 + 2w^3 - 10w + C$$

13. $\displaystyle\int \left(\frac{4}{x} + 5x^{-2} \right) dx = 4 \ln|x| + \frac{5x^{-1}}{-1} + C = 4 \ln|x| - \frac{5}{x} + C$

17. $\displaystyle\int (5 \sin x + 3 \cos x) \, dx = -5 \cos x + 3 \sin x + C$

21. $\displaystyle\int 15p^2 q^4 \, dp = 15 \left(\frac{p^3}{3} \right) q^4 + C = 5p^3 q^4 + C$

25. $\displaystyle\int 5e^{2q} \, dq = 5 \cdot \frac{1}{2} e^{2q} + C = 2.5 e^{2q} + C$

29. $\displaystyle\int (x^2 + 8 + e^x) \, dx = \frac{x^3}{3} + 8x + e^x + C$

33. $\displaystyle\int (Aq + B) \, dq = \frac{Aq^2}{2} + Bq + C$

37. $\displaystyle\int 12 \cos(4x) \, dx = 3 \sin(4x) + C$

41. We use the substitution $w = y + 2$, $dw = dy$:

$$\int \frac{1}{y+2} \, dy = \int \frac{1}{w} \, dw = \ln|w| + C = \ln|y+2| + C.$$

45. We use the substitution $w = 1 + \sin x$, $dw = \cos x \, dx$:

$$\int \frac{\cos x}{\sqrt{1 + \sin x}} \, dx = \int w^{-1/2} \, dw = \frac{w^{1/2}}{1/2} + C = 2\sqrt{1 + \sin x} + C.$$

CHAPTER EIGHT

Solutions for Section 8.1

1. Suppose x is the age of death; a possible density function is graphed in Figure 8.1.

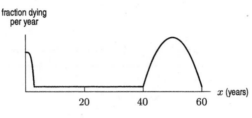

Figure 8.1

5. Since $p(x)$ is a density function, the area under the graph of $p(x)$ is 1, so

$$\text{Area} = \frac{1}{2}\text{Base} \cdot \text{Height} = \frac{1}{2} \cdot 10 \cdot a = 5a = 1$$

$$a = \frac{1}{5}.$$

9. We use the fact that the area of a triangle is $\frac{1}{2} \cdot \text{Base} \cdot \text{Height}$. Since $p(x)$ is a line with slope $0.1/20 = 0.005$, its equation is

$$p(x) = 0.005x.$$

(a) The fraction less than 5 meters high is the area to the left of 5. Since $p(5) = 0.005(5) = 0.025$,

$$\text{Fraction} = \frac{1}{2} \cdot 5(0.025) = 0.0625.$$

(b) The fraction more than 6 meters high is the area to the right of 6. Since $p(6) = 0.005(6) = 0.03$,

$$\text{Fraction} = 1 - (\text{Area to left of } 6)$$
$$= 1 - \frac{1}{2} \cdot 6(0.03) = 0.91.$$

(c) Fraction between 2 and 5 meters high is area between 2 and 5. Since $p(2) = 0.005(2) = 0.01$,

$$\text{Fraction} = (\text{Area to left of } 5) - (\text{Area to left of } 2)$$
$$= 0.0625 - \frac{1}{2} \cdot 2 \cdot (0.01) = 0.0525.$$

13. We can determine the fractions by estimating the area under the curve. Counting the squares for insects in the larval stage between 10 and 12 days we get 4.5 squares, with each square representing $(2) \cdot (3\%)$ giving a total of 27% of the insects in the larval stage between 10 and 12 days.

Likewise we get 2 squares for the insects in the larval stage for less than 8 days, giving 12% of the insects in the larval stage for less than 8 days.

Likewise we get 7.5 squares for the insects in the larval stage for more than 12 days, giving 45% of the insects in the larval stage for more than 12 days.

Since the peak of the graph occurs between 12 and 13 days, the length of the larval stage is most likely to fall in this interval.

17. See Figure 8.2. Many other answers are possible.

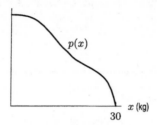

Figure 8.2

Solutions for Section 8.2

1. (a) See Figure 8.3. This is a cumulative distribution function.

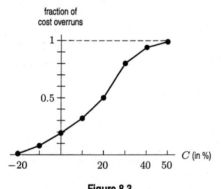

Figure 8.3

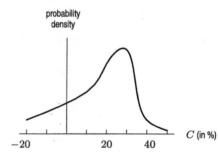

Figure 8.4

(b) The density function is the derivative of the cumulative distribution function. See Figure 8.4.
(c) Let's call the cumulative distribution function $F(C)$. The probability that there will be a cost overrun of more than 50% is $1 - F(50) = 0.01$, a 1% chance. The probability that it will be between 20% and 50% is $F(50) - F(20) = 0.99 - 0.50 = 0.49$, or 49%. The most likely amount of cost overrun occurs when the slope of the tangent line to the cumulative distribution function is a maximum. This occurs at the inflection point of the cumulative distribution graph, at about $C = 28\%$.

5.

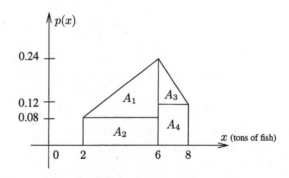

Splitting the figure into four pieces, we see that

$$\text{Area under the curve} = A_1 + A_2 + A_3 + A_4$$
$$= \frac{1}{2}(0.16)4 + 4(0.08) + \frac{1}{2}(0.12)2 + 2(0.12)$$
$$= 1.$$

We expect the area to be 1, since $\displaystyle\int_{-\infty}^{\infty} p(x)\,dx = 1$ for any probability density function, and $p(x)$ is 0 except when $2 \le x \le 8$.

9.

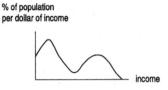

% of population
per dollar of income

Figure 8.5: Density function

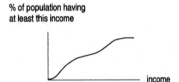

% of population having
at least this income

Figure 8.6: Cumulative distribution function

13.

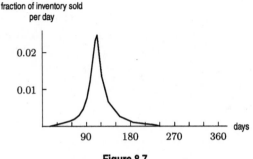

fraction of inventory sold
per day

Figure 8.7

17. **(a)** The probability that a banana lasts between 1 and 2 weeks is given by

$$\int_{1}^{2} p(t)dt = 0.25$$

Thus there is a 25% probability that the banana will last between one and two weeks.

(b) The formula given for $p(t)$ is valid for up to four weeks; for $t > 4$ we have $p(t) = 0$. So a banana lasting more than 3 weeks must last between 3 and 4 weeks. Thus the probability is

$$\int_{3}^{4} p(t)dt = 0.325$$

32.5% of the bananas last more than 3 weeks.

(c) Since $p(t) = 0$ for $t > 4$, the probability that a banana lasts more than 4 weeks is 0.

Solutions for Section 8.3

1. The median daily catch is the amount of fish such that half the time a boat will bring back more fish and half the time a boat will bring back less fish. Thus the area under the curve and to the left of the median must be 0.5. There are 25 squares under the curve so the median occurs at 12.5 squares of area. Now

$$\int_{2}^{5} p(x)dx = 10.5 \text{ squares}$$

and

$$\int_{5}^{6} p(x)dx = 5.5 \text{ squares,}$$

so the median occurs at a little over 5 tons. We must find the value a for which

$$\int_{5}^{a} p(t)dt = 2 \text{ squares,}$$

and we note that this occurs at about $a = 0.35$. Hence

$$\int_2^{5.35} p(t)\, dt \approx 12.5 \text{ squares}$$

$$\approx 0.5.$$

The median is about 5.35 tons.

5. (a) See Figure 8.8. The mean is larger than the median for this distribution; both are less than 15.

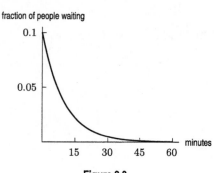

Figure 8.8

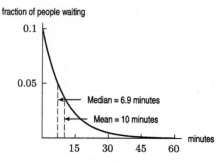

Figure 8.9

(b) We know that the median is the value T such that

$$\int_{-\infty}^{T} p(t)\,dt = 0.5$$

In our case this gives

$$0.5 = \int_0^T p(t)\,dt$$

$$= \int_0^T 0.1e^{-0.1t}\,dt$$

Substituting different values of T we get

$$\text{Median} = T \approx 6.9.$$

See Figure 8.9. We know that the mean is given by

$$\text{Mean} = \int_{-\infty}^{\infty} tp(t)\,dt$$

$$= \int_0^{60} (0.1te^{-0.1t})\,dt$$

$$\approx 9.83.$$

(c) The median tells us that exactly half of the people waiting at the stop wait less than 6.9 minutes.

The fact that the mean is 9.83 minutes can be interpreted in the following way: If all the people waiting at the stop were to wait exactly 9.83 minutes, the total time waited would be the same.

9. (a) Since $\mu = 100$ and $\sigma = 15$:

$$p(x) = \frac{1}{15\sqrt{2\pi}} e^{-\frac{1}{2}\left(\frac{x-100}{15}\right)^2}.$$

(b) The fraction of the population with IQ scores between 115 and 120 is (integrating numerically)

$$\int_{115}^{120} p(x)\, dx = \int_{115}^{120} \frac{1}{15\sqrt{2\pi}} e^{-\frac{(x-100)^2}{450}}\, dx$$

$$= \frac{1}{15\sqrt{2\pi}} \int_{115}^{120} e^{-\frac{(x-100)^2}{450}}\, dx$$

$$\approx 0.067 = 6.7\% \text{ of the population.}$$

Solutions for Chapter 8 Review

1. Since $p(x)$ is a density function,

$$\text{Area under graph} = \frac{1}{2} \cdot 50c = 25c = 1,$$

so $c = 1/25 = 0.04$.

5. For a small interval Δx around 68, the fraction of the population of American men with heights in this interval is about $(0.2)\Delta x$. For example, taking $\Delta x = 0.1$, we can say that approximately $(0.2)(0.1) = 0.02 = 2\%$ of American men have heights between 68 and 68.1 inches.

9. (a) The shaded region in Figure 8.10 represents the probability that the bus will be from 2 to 4 minutes late.

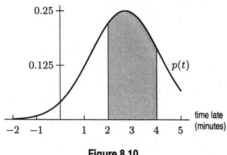

Figure 8.10

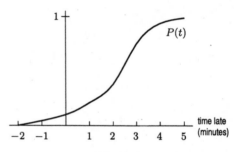

Figure 8.11

(b) The probability that the bus will be 2 to 4 minutes late (the area shaded in Figure 8.10) is $P(4) - P(2)$. The inflection point on the graph of $P(t)$ in Figure 8.11 corresponds to where $p(t)$ is a maximum. To the left of the inflection point, P is increasing at an increasing rate, while to the right of the inflection point P is increasing at a decreasing rate. Thus, the inflection point marks where the rate at which P is increasing is a maximum (i.e., where the derivative of P, which is p, is a maximum).

13. (a) Since $d(e^{-ct})/dt = ce^{-ct}$, we have

$$c\int_0^6 e^{-ct}\,dt = -e^{-ct}\Big|_0^6 = 1 - e^{-6c} = 0.1,$$

so

$$c = -\frac{1}{6}\ln 0.9 \approx 0.0176.$$

(b) Similarly, with $c = 0.0176$, we have

$$c\int_6^{12} e^{-ct}\,dt = -e^{-ct}\Big|_6^{12}$$
$$= e^{-6c} - e^{-12c} = 0.9 - 0.81 = 0.09,$$

so the probability is 9%.

17. True. The interval from $x = 9.98$ to $x = 10.04$ has length 0.06. Assuming that the value of $p(x)$ is near $1/2$ for $9.98 < x < 10.04$, the fraction of the population in that interval is $\int_{9.98}^{10.04} p(x)\,dx \approx (1/2)(0.06) = 0.03$.

CHAPTER NINE

Solutions for Section 9.1

1. Asking if f is an increasing or decreasing function of p is the same as asking how does f vary as we vary p, when we hold a fixed. Intuitively, we know that as we increase the price p, total sales of the product will go down. Thus, f is a decreasing function of p. Similarly, if we increase a, the amount spent on advertising, we can expect f to increase and therefore f is an increasing function of a.

5. (a) We expect B to be an increasing function of all three variables.
   (b) A deposit of $1250 at a 1% annual interest rate leads to a balance of $1276 after 25 months.

9. (a) According to Table 9.3 of the problem, it feels like $-19°F$.
   (b) A wind of 20 mph, according to Table 9.3.
   (c) About 17.5 mph. Since at a temperature of $25°F$, when the wind increases from 15 mph to 20 mph, the temperature adjusted for wind-chill decreases from $13°F$ to $11°F$, we can say that a 5 mph increase in wind speed causes an $2°F$ decrease in the temperature adjusted for wind-chill. Thus, each 2.5 mph increase in wind speed brings *about* a $1°F$ drop in the temperature adjusted for wind-chill. If the wind speed at $25°F$ increases from 15 mph to 17.5 mph, then the temperature you feel will be $13 - 1 = 12°F$.
   (d) Table 9.3 shows that with wind speed 20 mph the temperature will feel like $0°F$ when the air temperature is somewhere between $15°F$ and $20°F$. When the air temperature drops $5°F$ from $20°F$ to $15°F$, the temperature adjusted for wind-chill drops $6°F$ from $4°F$ to $-2°F$. We can say that for every $1°F$ decrease in air temperature there is *about* a $6/5 = 1.2°F$ drop in the temperature you feel. To drop the temperature you feel from $4°F$ to $0°F$ will take an air temperature drop of about $4/1.2 = 3.3°F$ from $20°F$. With a wind of 20 mph, approximately $20 - 3.3 = 16.7°F$ would feel like $0°F$."

13. Beef consumption by households making $20,000/year is given by Row 1 of Table 9.4 on page 347 of the text.

Table 9.1

p	3.00	3.50	4.00	4.50
$f(20, p)$	2.65	2.59	2.51	2.43

For households making $20,000/year, beef consumption decreases as price goes up.
Beef consumption by households making $100,000/year is given by Row 5 of Table 9.4.

Table 9.2

p	3.00	3.50	4.00	4.50
$f(100, p)$	5.79	5.77	5.60	5.53

For households making $100,000/year, beef consumption also decreases as price goes up.
Beef consumption by households when the price of beef is $3.00/lb is given by Column 1 of Table 9.4.

Table 9.3

I	20	40	60	80	100
$f(I, 3.00)$	2.65	4.14	5.11	5.35	5.79

When the price of beef is $3.00/lb, beef consumption increases as income increases.
Beef consumption by households when the price of beef is $4.00/lb is given by Column 3 of Table 9.4.

Table 9.4

I	20	40	60	80	100
$f(I, 4.00)$	2.51	3.94	4.97	5.19	5.60

When the price of beef is $4.00/lb, beef consumption increases as income increases.

17. In the answer to Problem 16 we saw that

$$P = 0.052\frac{M}{I},$$

and in the answer to Problem 15 we saw that

$$M = pf(I, p).$$

Putting the expression for M into the expression for P, gives:

$$P = 0.052\frac{pf(I, p)}{I}.$$

Solutions for Section 9.2

1. The temperature is decreasing away from the window, suggesting that heat is flowing in from the window. As time goes by the temperature at each point in the room increases. This could be caused by opening the window of an air conditioned room at $t = 0$ thus letting heat from the hot summer day outside raise the temperature inside.

5. Using our economic intuition, we know that the total sales of a product should be an increasing function of the amount spent on advertising. From the graph, Q is a decreasing function of x and an increasing function of y. Thus, the y-axis corresponds to the amount spent on advertising and the x-axis corresponds to the price of the product.

9. We first investigate the behavior of f with t fixed. We choose a value for t and move horizontally across the diagram looking at how the values on the contours change. For $t = 1$ hour, as we move from the left at $x = 0$ to the right at $x = 5$, we cross contours of 0.1, 0.2, 0.3, 0.4, and 0.5. The concentration of the drug is increasing as the initial dose, x, increases. (This is what we saw in Figure 9.2 in Section 9.1.) For each choice of f with t fixed (each horizontal line), we see that the contour values are increasing as we move from left to right, showing that, at any given time, the concentration of the drug increases as the size of the dose increases.

   The values of f with x fixed are read vertically. For $x = 4$, as we move up from $t = 0$ to $t = 5$, we see that the contours go up from 0.1 to 0.2 to 0.3, and then back down from 0.3 to 0.2 to 0.1. The maximum value of C is very close to 0.4 and is reached at about $t = 1$ hour. The concentration increases quickly at first (the contour lines are closer together), reaches its maximum, and then decreases slowly (the contour lines are farther apart.) All of the contours of f with x fixed are similar to this, although the maximum value varies. For the contour at $x = 3$, we see that the maximum value is slightly below 0.2. If the dose of the drug is 1 mg, the concentration of the drug in the bloodstream never gets as high as 0.1, since we cross no contours on the cross-section of f for $x = 1$.

13. To draw a contour for a wind-chill of $W = 20$, we need a few combinations of temperature and wind velocity (T, v) such that $W(T, v) = 20$. Estimating from the table, some such points are $(24, 5)$ and $(33, 10)$. We can connect these points to get a contour for $W = 20$. Similarly, some points that have wind-chill of about $0°F$ are $(5, 5)$, $(17.5, 10)$, $(23.5, 15)$, $(27, 20)$, and $(29, 25)$. By connecting these points we get the contour for $W = 0$. If we carry out this procedure for more values of W, we get a full contour diagram such as is shown in Figure 9.1:

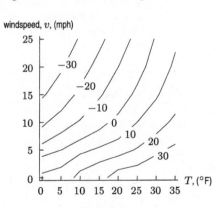

Figure 9.1

17. The contour where $f(x, y) = x + y + 1 = c$ or $y = -x + c - 1$ is the graph of the straight line of slope -1 as shown in Figure 9.2. Note that we have plotted the contours for $c = -2, -1, 0, 1, 2, 3, 4$. The contours are evenly spaced.

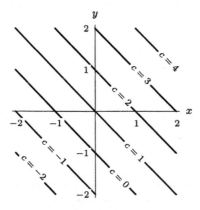

Figure 9.2

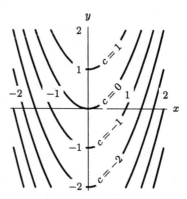

Figure 9.3

21. The contour where $f(x, y) = y - x^2 = c$ is the graph of the parabola $y = x^2 + c$, with vertex $(0, c)$ and symmetric about the y-axis, shown in Figure 9.3. Note that we have plotted the contours for $c = -2, -1, 0, 1$. The contours become more closely packed as we move farther from the y-axis.

25. (A) In graph I, $L = 1, K = 1$ gives us $F = 1$, and $L = 3, K = 3$ gives us $F = 3$. So tripling all inputs in graph I triples output; graph I corresponds to statement (A).

    (B) In graph II, $L = 1, K = 1$ gives us $F = 1$, and $L = 2.2, K = 2.2$ gives us $F = 1.5$. Extrapolating from this ratio, $L = 4, K = 4$ should gives us $F = 2$. So, quadrupling all inputs in graph II doubles output; graph II corresponds to statement (B).

    (C) In graph III, $L = 1, K = 1$ gives us $F = 1$, and $L = 2, K = 2$ gives us $F = 2.8$. So, doubling the inputs in graph III almost tripled output; graph III corresponds to statement (C).

29. (a) In this company success only increases when money increases, so success will remain constant along the work axis. However, as money increases so does success, which is shown in Graph (III).

    (b) As both work and money increase, success never increases, which corresponds to Graph (II).

    (c) If the money does not matter, then regardless of how much the money increases success will be constant along the money axis. However, success increases as work increases. This is best represented in Graph (I).

    (d) This company's success increases as both money and work increase, which is demonstrated in Graph (IV).

33. (a) If we have iron stomachs and can consume cola and pizza endlessly without ill effects, then we expect our happiness to increase without bound as we get more cola and pizza. Graph (IV) shows this since it increases along both the pizza and cola axes throughout.

    (b) If we get sick upon eating too many pizzas or drinking too much cola, then we expect our happiness to decrease once either or both of those quantities grows past some optimum value. This is depicted in graph (I) which increases along both axes until a peak is reached, and then decreases along both axes.

    (c) If we do get sick after too much cola, but are always able to eat more pizza, then we expect our happiness to decrease after we drink some optimum amount of cola, but continue to increase as we get more pizza. This is shown by graph (III) which increases continuously along the pizza axis but, after reaching a maximum, begins to decrease along the cola axis.

37. (a) For $g(x, t) = \cos 2t \sin x$, our snapshots for fixed values of t are still one arch of the sine curve. The amplitudes, which are governed by the $\cos 2t$ factor, now change twice as fast as before. That is, the string is vibrating twice as fast.

    (b) For $y = h(x, t) = \cos t \sin 2x$, the vibration of the string is more complicated. If we hold t fixed at any value, the snapshot now shows one full period, i.e. one crest and one trough, of the sine curve. The magnitude of the sine curve is time dependent, given by $\cos t$. Now the center of the string, $x = \pi/2$, remains stationary just like the end points. This is a vibrating string with the center held fixed, as shown in Figure 9.4.

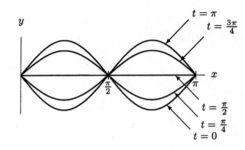

Figure 9.4: Another vibrating string: $y = h(x, t) = \cos t \sin 2x$

Solutions for Section 9.3

1. (a) Positive.
   (b) Negative.
   (c) Positive.
   (d) Zero.

5. (a) We expect the demand for coffee to decrease as the price of coffee increases (assuming the price of tea is fixed.) Thus we expect f_c to be negative. We expect people to switch to coffee as the price of tea increases (assuming the price of coffee is fixed), so that the demand for coffee will increase. We expect f_t to be positive.
   (b) The statement $f(3, 2) = 780$ tells us that if coffee costs \$3 per pound and tea costs \$2 per pound, we can expect 780 pounds of coffee to sell each week. The statement $f_c(3, 2) = -60$ tells us that, if the price of coffee then goes up \$1 and the price of tea stays the same, the demand for coffee will go down by about 60 pounds. The statement $20 = f_t(3, 2)$ tells us that if the price of tea goes up \$1 and the price of coffee stays the same, the demand for coffee will go up by about 20 pounds.

9. (a) We expect f_p to be negative because if the price of the product increases, the sales usually decrease.
   (b) If the price of the product is \$8 per unit and if \$12000 has been spent on advertising, sales increase by approximately 150 units if an additional \$1000 is spent on advertising.

13. (a) Since $f_x > 0$, the values on the contours increase as you move to the right. Since $f_y > 0$, the values on the contours increase as you move upward. See Figure 9.5.

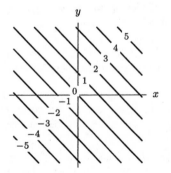

Figure 9.5: $f_x > 0$ and $f_y > 0$

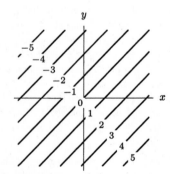

Figure 9.6: $f_x > 0$ and $f_y < 0$

(b) Since $f_x > 0$, the values on the contours increase as you move to the right. Since $f_y < 0$, the values on the contours decrease as you move upward. See Figure 9.6.
(c) Since $f_x < 0$, the values on the contours decrease as you move to the right. Since $f_y > 0$, the values on the contours increase as you move upward. See Figure 9.7.

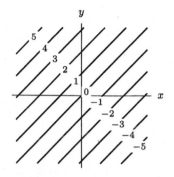

Figure 9.7: $f_x < 0$ and $f_y > 0$

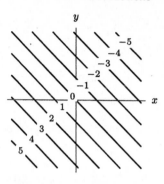

Figure 9.8: $f_x < 0$ and $f_y < 0$

(d) Since $f_x < 0$, the values on the contours decrease as you move to the right. Since $f_y < 0$, the values on the contours decrease as you move upward. See Figure 9.8.

17. We can use the formula $\Delta f \approx \Delta r f_r + \Delta s f_s$. Applying this formula in order to estimate $f(52, 108)$ from the known value of $f(50, 100)$ gives

$$f(52, 108) \approx f(50, 100) + (52 - 50)f_r(50, 100) + (108 - 100)f_s(50, 100)$$
$$= 5.67 + 2(0.60) + 8(-0.15)$$
$$= 5.67.$$

21. For $f_w(10, 25)$ we get

$$f_w(10, 25) \approx \frac{f(10 + h, 25) - f(10, 25)}{h}.$$

Choosing $h = 5$ and reading values from Table 9.3 on page 347 of the text, we get

$$f_w(10, 25) \approx \frac{f(15, 25) - f(10, 25)}{5} = \frac{13 - 15}{5} = -0.4°\text{F/mph}$$

This means that when the wind speed is 10 mph and the true temperature is $25°$F, as the wind speed increases from 10 mph by 1 mph we feel an approximately $0.4°$F drop in temperature. This rate is negative because the temperature you feel drops as the wind speed increases.

25. (a) The partial derivative $f_x = 350$ tells us that R increases by \$350 as x increases by 1. Thus, $f(201, 400) = f(200, 400) + 350 = 150{,}000 + 350 = 150{,}350$.
  (b) The partial derivative $f_y = 200$ tells us that R increases by \$200 as y increases by 1. Since y is increasing by 5, we have $f(200, 405) = f(200, 400) + 5(200) = 150{,}000 + 1{,}000 = 151{,}000$.
  (c) Here x is increasing by 3 and y is increasing by 6. We have $f(203, 406) = f(200, 400) + 3(350) + 6(200) = 150{,}000 + 1{,}050 + 1{,}200 = 152{,}250$.
  In this problem, the partial derivatives gave exact results, but in general they only give an estimate of the changes in the function.

Solutions for Section 9.4

1. $f_x(x, y) = 2x + 2y$, $f_y(x, y) = 2x + 3y^2$.

5. $\dfrac{\partial P}{\partial r} = 100te^{rt}$

9. $z_x = 2xy + 10x^4 y$

13. $f_x(x, y) = 10xy^3 + 8y^2 - 6x$ and $f_y(x, y) = 15x^2 y^2 + 16xy$.

17.

$$f(3, 1) = 5(3)(1)^2 = 15$$
$$f_u(u, v) = 5v^2 \Rightarrow f_u(3, 1) = 5(1)^2 = 5$$
$$f_v(u, v) = 10uv \Rightarrow f_v(3, 1) = 10(3)(1) = 30$$

21. Substituting $w = 65$ and $h = 160$, we have

$$f(65, 160) = 0.01(65^{0.25})(160^{0.75}) = 1.277 \text{ m}^2.$$

This tells us that a person who weighs 65 kg and is 160 cm tall has a surface area of about 1.277 m^2. Since

$$f_w(w, h) = 0.01(0.25w^{-0.75})h^{0.75} \text{ m}^2/\text{kg},$$

we have $f_w(65, 160) = 0.005 \text{ m}^2/\text{kg}$. Thus, an increase of 1 kg in weight increases surface area by about 0.005 m^2. Since

$$f_h(w, h) = 0.01w^{0.25}(0.75h^{-0.25}) \text{ m}^2/\text{cm},$$

we have $f_h(65, 160) = 0.006 \text{ m}^2/\text{cm}$. Thus, an increase of 1 cm in height increases surface area by about 0.006 m^2.

25. (a) At Q, R, we have $f_x < 0$ because f decreases as we move in the x-direction.
   (b) At Q, P, we have $f_y > 0$ because f increases as we move in the y-direction.
   (c) At all four points, P, Q, R, S, we have $f_{xx} > 0$, because f_x is increasing as we move in the x-direction. (At P, S, we see that f_x is positive and getting larger; at Q, R, we see that f_x is negative and getting less negative.)
   (d) At all four points, P, Q, R, S, we have $f_{yy} > 0$, so there are none with $f_{yy} < 0$. The reasoning is similar to part (c).

29. $f_x = 2/y$ and $f_y = -2x/y^2$, so $f_{xx} = 0$, $f_{xy} = -2/y^2$, $f_{yy} = 4x/y^3$ and $f_{yx} = -2/y^2$.

33. $V_r = 2\pi rh$ and $V_h = \pi r^2$, so $V_{rr} = 2\pi h$, $V_{hh} = 0$ and $V_{rh} = V_{hr} = 2\pi r$.

37. $f_r = 100te^{rt}$ and $f_t = 100re^{rt}$, so $f_{rr} = 100t^2e^{rt}$, $f_{rt} = f_{tr} = 100tre^{rt} + 100e^{rt} = 100(rt + 1)e^{rt}$ and $f_{tt} = 100r^2e^{rt}$.

Solutions for Section 9.5

1. We can identify local extreme points on a contour diagram because these points will either be the centers of a series of concentric circles that close around them, or will lie on the edges of the diagram. Looking at the graph, we see that $(2, 10)$, $(6, 4)$, $(6.5, 16)$ and $(9, 10)$ appear to be such points. Since the points near $(2, 10)$ decrease in functional value as they close around $(2, 10)$, $f(2, 10)$ will be somewhat less than its nearest contour. So $f(2, 10) \approx 0.5$. Similarly, since the contours near $(2, 10)$ are greater in functional value than $f(2, 10)$, $f(2, 10)$ is a local minimum. Applying analogous arguments to the point $(6, 4)$, we see that $f(6, 4) \approx 9.5$ and is a local maximum. The contour values are increasing as we approach $(6.5, 16)$ along any path, so $f(6.5, 16) \approx 10$ is a local maximum and $(9, 10)$ is a local minimum.
   Since none of the local minima are less in value than $f(2, 10) \approx 0.5$, $f(2, 10)$ is a global minimum. Since none of the local maxima are greater in value than $f(6.5, 16) \approx 10$, $f(6.5, 16)$ is a global maximum.

5. At a critical point $f_x = 2x + y = 0$ and $f_y = x + 3 = 0$, so $(-3, 6)$ is the only critical point. Since $f_{xx}f_{yy} - f_{xy}^2 = -1 < 0$, the point $(-3, 6)$ is neither a local maximum nor a local minimum.

9. To find the critical points, we solve $f_x = 0$ and $f_y = 0$ for x and y. Solving

$$f_x = 2x - 2y = 0,$$
$$f_y = -2x + 6y - 8 = 0.$$

We see from the first equation that $x = y$. Substituting this into the second equation shows that $y = 2$. The only critical point is $(2, 2)$.
   We have

$$D = (f_{xx})(f_{yy}) - (f_{xy})^2 = (2)(6) - (-2)^2 = 8.$$

Since $D > 0$ and $f_{xx} = 2 > 0$, the function f has a local minimum at the point $(2, 2)$.

13. Mississippi lies entirely within a region designated as 80s so we expect both the maximum and minimum daily high temperatures within the state to be in the 80s. The southwestern-most corner of the state is close to a region designated as 90s, so we would expect the temperature here to be in the high 80s, say 87-88. The northern-most portion of the state is located near the center of the 80s region. We might expect the high temperature there to be between 83-87.
   Alabama also lies completely within a region designated as 80s so both the high and low daily high temperatures within the state are in the 80s. The southeastern tip of the state is close to a 90s region so we would expect the temperature here to be about 88-89 degrees. The northern-most part of the state is near the center of the 80s region so the temperature there is 83-87 degrees.

Pennsylvania is also in the 80s region, but it is touched by the boundary line between the 80s and a 70s region. Thus we expect the low daily high temperature to occur there and be about 80 degrees. The state is also touched by a boundary line of a 90s region so the high will occur there and be 89-90 degrees.

New York is split by a boundary between an 80s and a 70s region, so the northern portion of the state is likely to be about 74-76 while the southern portion is likely to be in the low 80s, maybe 81-84 or so.

California contains many different zones. The northern coastal areas will probably have the daily high as low as 65-68, although without another contour on that side, it is difficult to judge how quickly the temperature is dropping off to the west. The tip of Southern California is in a 100s region, so there we expect the daily high to be 100-101.

Arizona will have a low daily high around 85-87 in the northwest corner and a high in the 100s, perhaps 102-107 in its southern regions.

Massachusetts will probably have a high daily high around 81-84 and a low daily high of 70.

17. At a local maximum value of f,

$$\frac{\partial f}{\partial x} = -2x - B = 0.$$

We are told that this is satisfied by $x = -2$. So $-2(-2) - B = 0$ and $B = 4$. In addition,

$$\frac{\partial f}{\partial y} = -2y - C = 0$$

and we know this holds for $y = 1$, so $-2(1) - C = 0$, giving $C = -2$. We are also told that the value of f is 15 at the point $(-2, 1)$, so

$$15 = f(-2, 1) = A - ((-2)^2 + 4(-2) + 1^2 - 2(1)) = A - (-5), \text{ so } A = 10.$$

Now we check that these values of A, B, and C give $f(x, y)$ a local maximum at the point $(-2, 1)$. Since

$$f_{xx}(-2, 1) = -2,$$
$$f_{yy}(-2, 1) = -2$$

and

$$f_{xy}(-2, 1) = 0,$$

we have that $f_{xx}(-2, 1)f_{yy}(-2, 1) - f_{xy}^2(-2, 1) = (-2)(-2) - 0 > 0$ and $f_{xx}(-2, 1) < 0$. Thus, f has a local maximum value 15 at $(-2, 1)$.

21. The total revenue is

$$R = pq = (60 - 0.04q)q = 60q - 0.04q^2,$$

and as $q = q_1 + q_2$, this gives

$$R = 60q_1 + 60q_2 - 0.04q_1^2 - 0.08q_1q_2 - 0.04q_2^2.$$

Therefore, the profit is

$$P(q_1, q_2) = R - C_1 - C_2$$
$$= -13.7 + 60q_1 + 60q_2 - 0.07q_1^2 - 0.08q_2^2 - 0.08q_1q_2.$$

At a local maximum point, we have:

$$\frac{\partial P}{\partial q_1} = 60 - 0.14q_1 - 0.08q_2 = 0,$$

$$\frac{\partial P}{\partial q_2} = 60 - 0.16q_2 - 0.08q_1 = 0.$$

Solving these equations, we find that

$$q_1 = 300 \quad \text{and} \quad q_2 = 225.$$

To see whether or not we have found a local maximum, we compute the second-order partial derivatives:

$$\frac{\partial^2 P}{\partial q_1^2} = -0.14, \quad \frac{\partial^2 P}{\partial q_2^2} = -0.16, \quad \frac{\partial^2 P}{\partial q_1 \partial q_2} = -0.08.$$

Therefore,

$$D = \frac{\partial^2 P}{\partial q_1^2}\frac{\partial^2 P}{\partial q_2^2} - \frac{\partial^2 P}{\partial q_1 \partial q_2} = (-0.14)(-0.16) - (-0.08)^2 = 0.016,$$

and so we have found a local maximum point. The graph of $P(q_1, q_2)$ has the shape of an upside down paraboloid since P is quadratic in q_1 and q_2, hence $(300, 225)$ is a global maximum point.

Solutions for Section 9.6

1. We wish to optimize $f(x, y) = xy$ subject to the constraint $g(x, y) = 5x + 2y = 100$. To do this we must solve the following system of equations:

$$f_x(x, y) = \lambda g_x(x, y), \qquad \text{so } y = 5\lambda$$
$$f_y(x, y) = \lambda g_y(x, y), \qquad \text{so } x = 2\lambda$$
$$g(x, y) = 100, \qquad \text{so } 5x + 2y = 100$$

We substitute in the third equation to obtain $5(2\lambda) + 2(5\lambda) = 100$, so $\lambda = 5$. Thus,

$$x = 10 \quad y = 25 \quad \lambda = 5$$

corresponding to optimal $f(x, y) = (10)(25) = 250$.

5. Our objective function is $f(x, y) = x + y$ and our equation of constraint is $g(x, y) = x^2 + y^2 = 1$. To optimize $f(x, y)$ with Lagrange multipliers, we solve the following system of equations

$$f_x(x, y) = \lambda g_x(x, y), \qquad \text{so } 1 = 2\lambda x$$
$$f_y(x, y) = \lambda g_y(x, y), \qquad \text{so } 1 = 2\lambda y$$
$$g(x, y) = 1, \qquad \text{so } x^2 + y^2 = 1$$

Solving for λ gives

$$\lambda = \frac{1}{2x} = \frac{1}{2y},$$

which tells us that $x = y$. Going back to our equation of constraint, we use the substitution $x = y$ to solve for y:

$$g(y, y) = y^2 + y^2 = 1$$
$$2y^2 = 1$$
$$y^2 = \frac{1}{2}$$
$$y = \pm\sqrt{\frac{1}{2}} = \pm\frac{\sqrt{2}}{2}.$$

Since $x = y$, our critical points are $(\frac{\sqrt{2}}{2}, \frac{\sqrt{2}}{2})$ and $(-\frac{\sqrt{2}}{2}, -\frac{\sqrt{2}}{2})$. Since the constraint is closed and bounded, maximum and minimum values of f subject to the constraint exist. Evaluating f at the critical points we find that the maximum value is $f(\frac{\sqrt{2}}{2}, \frac{\sqrt{2}}{2}) = \sqrt{2}$ and the minimum value is $f(-\frac{\sqrt{2}}{2}, -\frac{\sqrt{2}}{2}) = -\sqrt{2}$.

9. Our objective function is $f(x, y) = xy$ and our equation of constraint is $g(x, y) = 4x^2 + y^2 = 8$. Their partial derivatives are

$$f_x = y, \qquad f_y = x$$
$$g_x = 8x, \qquad g_y = 2y.$$

This gives

$$8x\lambda = y \quad \text{and} \quad 2y\lambda = x.$$

Multiplying, we get

$$8x^2\lambda = 2y^2\lambda.$$

If $\lambda = 0$, then $x = y = 0$, which does not satisfy the constraint equation. So $\lambda \neq 0$ and we get

$$2y^2 = 8x^2$$
$$y^2 = 4x^2$$
$$y = \pm 2x.$$

To find x, we substitute for y in our equation of constraint.

$$4x^2 + y^2 = 8$$
$$4x^2 + 4x^2 = 8$$
$$x^2 = 1$$
$$x = \pm 1$$

So our critical points are $(1,2)$, $(1,-2)$, $(-1,2)$ and $(-1,-2)$. Evaluating $f(x,y)$ at the critical points, we have

$$f(1,2) = f(-1,-2) = 2$$
$$f(1,-2) = f(1,-2) = -2.$$

Thus, the maximum value of f on $g(x,y) = 8$ is 2, and the minimum value is -2.

13. **(a)** Objective function: $Q = x_1^{0.3} x_2^{0.7}$.
   (b) Constraint: $10x_1 + 25x_2 = 50{,}000$.

17. We want to minimize

$$C = f(q_1, q_2) = 2q_1^2 + q_1 q_2 + q_2^2 + 500$$

subject to the constraint $q_1 + q_2 = 200$ or $g(q_1, q_2) = q_1 + q_2 = 200$.
   We solve the system of equations:

$$C_{q_1} = \lambda g_{q_1}, \qquad \text{so } 4q_1 + q_2 = \lambda$$
$$C_{q_2} = \lambda g_{q_2}, \qquad \text{so } 2q_2 + q_1 = \lambda$$
$$g = 200, \qquad \text{so } q_1 + q_2 = 200.$$

Solving we get

$$4q_1 + q_2 = 2q_2 + q_1$$

so

$$3q_1 = q_2.$$

We want

$$q_1 + q_2 = 200$$
$$q_1 + 3q_1 = 4q_1 = 200.$$

Therefore

$$q_1 = 50 \text{ units}, \quad q_2 = 150 \text{ units}.$$

21. **(a)** The objective function is the function that is optimized. Since the problem refers to maximum production, the objective function is the production function $P(K, L)$.
   (b) Production is maximized subject to a budget restriction, which is the constraint. The constraint equation is $C(K, L) = 600{,}000$.
   (c) The Lagrange multiplier tells you the rate at which maximum production changes when the budget is increased. Its units are tons of steel per dollar of budget, or simply tons/dollar.
   (d) Increasing the budget from \$600,000 to \$(600,000+a) increases the maximum possible production from 2,500,000 tons to approximately $(2{,}500{,}000 + 3.17a)$ tons. Every extra dollar of budget increases maximal production by approximately $\lambda = 3.17$ tons.

25. The largest value of $f(x, y)$ on the line $y = 100$ shown in Figure 9.9 occurs at the left endpoint, at the point $(0, 100)$. The maximum value is $f(0, 100) \approx 43$.

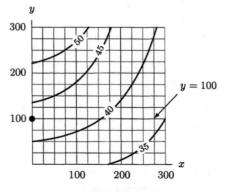

Figure 9.9

29. (a) We have 1500 workers and $4,000,000 per month of capital, so $x = 1500, y = 4,000,000/1000 = 4000$. Substituting into the equation for Q gives us $Q = (1500)^{0.4}(4000)^{0.6} = 2702$ cars per month.

(b) Now we are only producing 2000 cars per month. We wish to minimize cost subject to the constraint that the monthly production is 2000 cars. So, our objective function is cost $C = 2100x + 1000y$ and our constraint is that $Q = x^{0.4}y^{0.6} = 2000$. To minimize C according to this, we solve the following system of equations:

$$C_x = \lambda Q_x, \qquad \text{so } 0.4x^{-0.6}y^{0.6}\lambda = 2100$$
$$C_y = \lambda Q_x, \qquad \text{so } 0.6x^{0.4}y^{-0.4}\lambda = 1000$$
$$Q = 2000, \qquad \text{so } x^{0.4}y^{0.6} = 2000$$

Dividing the first two equations gives

$$\frac{0.4x^{-0.6}y^{0.6}\lambda}{0.6x^{0.4}y^{-0.4}\lambda} = \frac{0.4}{0.6}\frac{y}{x} = \frac{2100}{1000} \Rightarrow y = 3.15x.$$

Substituting this into the constraint gives us $x \approx 1004.72$ and $y \approx 3164.858$. So our new level of production uses 1005 workers and $3,164,858 of equipment. So $1500 - 1005 = 495$ workers will be laid off, and monthly investment in capital will fall by $4,000,000 - $3,164,858 = $835,142.

(c) Solving for the Lagrange multiplier λ from the above equations gives us

$$\lambda = \frac{2100}{0.4x^{-0.6}y^{0.6}} = \frac{2100}{0.4(1004.72)^{-0.6}(3164.86)^{0.6}} \approx \$2637.4 \text{ per car}$$

This means that to produce on additional car per month would cost about $2637.40 with the lowest-cost use of capital and labor.

33. (a) The curves are shown in Figure 9.10.

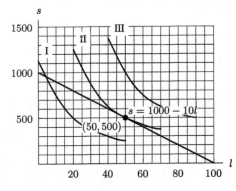

Figure 9.10

(b) The income equals $10/hour times the number of hours of work:

$$s = 10(100 - l) = 1000 - 10l.$$

(c) The graph of this constraint is the straight line in Figure 9.10.

(d) For any given salary, curve III allows for the most leisure time, curve I the least. Similarly, for any amount of leisure time, curve III also has the greatest salary, and curve I the least. Thus, any point on curve III is preferable to any point on curve II, which is preferable to any point on curve I. We prefer to be on the outermost curve that our constraint allows. We want to choose the point on $s = 1000 - 10l$ which is on the most preferable curve. Since all the curves are concave up, this occurs at the point where $s = 1000 - 10l$ is *tangent* to curve II. So we choose $l = 50, s = 500$, and work 50 hours a week.

Solutions for Chapter 9 Review

1. (a) It feels like $81°F$.

(b) At 30% relative humidity, 90°F feels like 90°F.

(c) By finding the temperature which has heat index 105°F for each humidity level, we get Table 9.5:

Table 9.5 *Estimates of danger temperatures*

Relative humidity(%)	0	10	20	30	40	50	60
Temperature(°F)	117	110	105	101	97	94	92

(d) With a high humidity your body cannot cool itself as well by sweating. With a low humidity your body is capable of cooling itself to below the actual temperature. Therefore a high humidity feels hotter and a low humidity feels cooler.

5. (a)

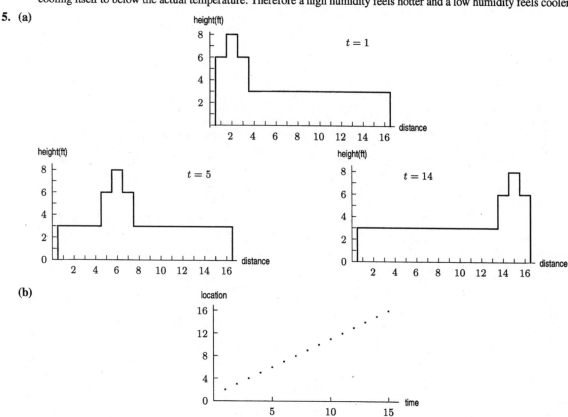

(b)

Figure 9.11

(c) The "wave" at a sports arena.

9. The point $x = 10$, $t = 5$ is between the contours $H = 70$ and $H = 75$, a little closer to the former. Therefore, we estimate $H(10, 5) \approx 72$, i.e., it is about 72°F. Five minutes later we are at the point $x = 10$, $t = 10$, which is just above the contour $H = 75$, so we estimate that it has warmed up to 76°F by then.

13. Q is a decreasing function of c and an increasing function of t. This is because when the price of coffee rises, consumers drink less. When the price of tea rises, some consumers switch from tea to coffee and the demand for coffee increases.

17. (a) The TMS map of an eye of constant curvature will have only one color, with no contour lines dividing the map.

(b) The contour lines are circles, because the cross-section is the same in every direction. The largest curvature is in the center. See picture below.

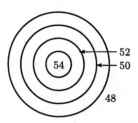

21. We estimate $\partial I/\partial H$ and $\partial I/\partial T$ by using difference quotients. We have

$$\frac{\partial I}{\partial H} \approx \frac{f(H + \Delta H, T) - f(H,T)}{\Delta H} \quad \text{and} \quad \frac{\partial I}{\partial T} \approx \frac{f(H, T + \Delta T) - f(H,T)}{\Delta T}$$

Choosing $\Delta H = 10$ and reading the values from Table 9.5 on page 353 in the text we get

$$\left.\frac{\partial I}{\partial H}\right|_{(10,100)} \approx \frac{f(10 + 10, 100) - f(10, 100)}{10} = \frac{99 - 95}{10} = 0.4.$$

Similarly, choosing $\Delta T = 5$ we get

$$\left.\frac{\partial I}{\partial T}\right|_{(10,100)} \approx \frac{f(10, 100 + 5) - f(10, 100)}{5} = \frac{100 - 95}{5} = 1.$$

The fact that $\left.\dfrac{\partial I}{\partial H}\right|_{(10,100)} \approx 0.4$ means that the rate of change of the heat index per unit increase in humidity is about 0.4. This means that the heat index increases by approximately $0.4°F$ for every percentage point increase in humidity. This rate is positive, because as the humidity increases, the heat index increases.

The partial derivative $\partial I/\partial T$ gives the rate of increase of heat index with respect to temperature. It is positive because the heat index increases as temperature increases. Knowing that $\left.\dfrac{\partial I}{\partial T}\right|_{(10,100)} \approx 1$ tells us that as the temperature increases by $1°F$, the temperature you feel increases by $1°F$ also. In other words, at this humidity and temperature, the changes in temperature that you feel are approximately equal to the actual changes in temperature.

The fact that $\dfrac{\partial I}{\partial T} > \dfrac{\partial I}{\partial H}$ at $(10, 100)$ tells us that a unit increase in temperature has a greater effect on the heat index in Tucson than a unit increase in humidity.

25. Estimating from the contour diagram, using positive increments for Δx and Δy, we have, for point A,

$$\left.\frac{\partial n}{\partial x}\right|_{(A)} \approx \frac{1.5 - 1}{67 - 59} = \frac{1/2}{8} = \frac{1}{16} \approx 0.06 \quad \frac{\text{foxes/km}^2}{\text{km}}$$

$$\left.\frac{\partial n}{\partial y}\right|_{(A)} \approx \frac{0.5 - 1}{60 - 51} = -\frac{1/2}{9} = -\frac{1}{18} \approx -0.06 \quad \frac{\text{foxes/km}^2}{\text{km}}.$$

So, from point A the fox population density increases as we move eastward. The population density decreases as we move north from A.

At point B,

$$\left.\frac{\partial n}{\partial x}\right|_{(B)} \approx \frac{0.75 - 1}{135 - 115} = -\frac{1/4}{20} = -\frac{1}{80} \approx -0.01 \quad \frac{\text{foxes/km}^2}{\text{km}}$$

$$\left.\frac{\partial n}{\partial y}\right|_{(B)} \approx \frac{0.5 - 1}{120 - 110} = -\frac{1/2}{10} = -\frac{1}{20} \approx -0.05 \quad \frac{\text{foxes/km}^2}{\text{km}}.$$

So, fox population density decreases as we move both east and north of B. However, notice that the partial derivative $\partial n/\partial x$ at B is smaller in magnitude than the others. Indeed if we had taken a negative Δx we would have obtained an estimate of the opposite sign. This suggests that better estimates for B are

$$\left.\frac{\partial n}{\partial x}\right|_{(B)} \approx 0 \quad \frac{\text{foxes/km}^2}{\text{km}}$$

$$\left.\frac{\partial n}{\partial y}\right|_{(B)} \approx -0.05 \quad \frac{\text{foxes/km}^2}{\text{km}}.$$

At point C,

$$\left.\frac{\partial n}{\partial x}\right|_{(C)} \approx \frac{2 - 1.5}{135 - 115} = \frac{1/2}{20} = \frac{1}{40} \approx 0.02 \quad \frac{\text{foxes/km}^2}{\text{km}}$$

$$\left.\frac{\partial n}{\partial y}\right|_{(C)} \approx \frac{2 - 1.5}{80 - 55} = \frac{1/2}{25} = \frac{1}{50} \approx 0.02 \quad \frac{\text{foxes/km}^2}{\text{km}}.$$

So, the fox population density increases as we move east and north of C. Again, if these estimates were made using negative values for Δx and Δy we would have had estimates of the opposite sign. Thus, better estimates are

$$\left.\frac{\partial n}{\partial x}\right|_{(C)} \approx 0 \quad \frac{\text{foxes/km}^2}{\text{km}}$$

$$\left.\frac{\partial n}{\partial y}\right|_{(C)} \approx 0 \quad \frac{\text{foxes/km}^2}{\text{km}}.$$

29. $P_a = 2a - 2b^2$, $P_b = -4ab$.

33. $f_x = \frac{x}{\sqrt{x^2+y^2}}$, $f_y = \frac{y}{\sqrt{x^2+y^2}}$.

37. Since $\dfrac{\partial f}{\partial y}$ and $\dfrac{\partial f}{\partial y}$ are defined everywhere, a critical point will occur where $\dfrac{\partial f}{\partial x} = 0$ and $\dfrac{\partial f}{\partial y} = 0$. So:

$$\frac{\partial f}{\partial x} = 2x - 4 = 0 \Rightarrow x = 2$$

$$\frac{\partial f}{\partial y} = 6y + 6 = 0 \Rightarrow y = -1$$

$(2, -1)$ is the critical point of $f(x, y)$.

41. (a) We wish to maximize

$$V = 1000D^{0.6}N^{0.3}$$

subject to the budget constraint in dollars

$$40{,}000D + 10{,}000N \leq 600{,}000$$

or (in thousand dollars)

$$\text{Cost } C = 40D + 10N \leq 600.$$

(b) Since additional doctors and nurses will always increase visits, we will use the entire budget allotted to us, so $C = 40D + 10N = 600$. To optimize V subject to this constraint, we must solve the following system of equations:

$$V_D = \lambda C_D, \text{ so } 600D^{-0.4}N^{0.3} = 40\lambda$$
$$V_N = \lambda C_N, \text{ so } 300D^{0.6}N^{-0.7} = 10\lambda$$
$$C = 600, \text{ so } 40D + 10N = 600$$

Thus, we get

$$\frac{600D^{-0.4}N^{0.3}}{40} = \lambda = \frac{300D^{0.6}N^{-0.7}}{10}$$

So

$$N = 2D.$$

To solve for D and N, substitute in the budget constraint:

$$40D + 10(2D) = 600$$

$$60D = 600$$

So $D = 10$ and $N = 20$.

$$\lambda = \frac{600(10^{-0.4})(20^{0.3})}{40} \approx 14.67$$

Thus the clinic should hire 10 doctors and 20 nurses. With that staff, the clinic can provide

$$V = 1000(10^{0.6})(20^{0.3}) \approx 9{,}779 \text{ visits per year.}$$

(c) From part (b), the Lagrange multiplier is $\lambda = 14.67$. At the optimum, the Lagrange multiplier tells us that about 14.67 extra visits can be generated through an increase of \$1,000 in the budget. (If we had written out the constraint in dollars instead of thousands of dollars, the Lagrange multiplier would tell us the number of extra visits per dollar.)

Solutions to Problems on Deriving the Formula for Regression Lines ————————

1. Let the line be in the form $y = b + mx$. When x equals -1, 0 and 1, then y equals $b - m$, b, and $b + m$, respectively. The sum of the squares of the vertical distances, which is what we want to minimize, is

$$f(m, b) = (2 - (b - m))^2 + (-1 - b)^2 + (1 - (b + m))^2.$$

To find the critical points, we compute the partial derivatives with respect to m and b,

$$\begin{aligned}
f_m &= 2(2 - b + m) + 0 + 2(1 - b - m)(-1) \\
&= 4 - 2b + 2m - 2 + 2b + 2m \\
&= 2 + 4m, \\
f_b &= 2(2 - b + m)(-1) + 2(-1 - b)(-1) + 2(1 - b - m)(-1) \\
&= -4 + 2b - 2m + 2 + 2b - 2 + 2b + 2m \\
&= -4 + 6b.
\end{aligned}$$

Setting both partial derivatives equal to zero, we get a system of equations:

$$2 + 4m = 0,$$
$$-4 + 6b = 0.$$

The solution is $m = -1/2$ and $b = 2/3$. You can check that it is a minimum. Hence, the regression line is $y = \dfrac{2}{3} - \dfrac{1}{2}x$.

5. We have $\sum x_i = 6$, $\sum y_i = 5$, $\sum x_i^2 = 14$, and $\sum y_i x_i = 12$. Thus

$$b = (14 \cdot 5 - 6 \cdot 12) / \left(3 \cdot 14 - 6^2\right) = -1/3.$$
$$m = (3 \cdot 12 - 6 \cdot 5) / \left(3 \cdot 14 - 6^2\right) = 1.$$

The line is $y = x - \dfrac{1}{3}$, which agrees with the answer to Example 1.

CHAPTER TEN

Solutions for Section 10.1

1. The rate of change of P is proportional to P so we have

$$\frac{dP}{dt} = kP,$$

for some constant k. Since the population P is increasing, the derivative dP/dt must be positive. Therefore, k is positive.

5. (a) = (III), (b) = (IV), (c) = (I), (d) = (II).

9. The amount of morphine, M, is increasing at a rate of 2.5 mg/hour and is decreasing at a rate of 0.347 times M. We have

Rate of change of M = Rate in − Rate out.

$$\frac{dM}{dt} = 2.5 - 0.347M.$$

13. (a) = (I), (b) = (IV), (c) = (III). Graph (II) represents an egg originally at 0° C which is moved to the kitchen table (20° C) two minutes after the egg in part (a) is moved.

Solutions for Section 10.2

1. Since $y = t^4$, the derivative is $dy/dt = 4t^3$. We have

$$\text{Left-side} = t\frac{dy}{dt} = t(4t^3) = 4t^4.$$

$$\text{Right-side} = 4y = 4t^4.$$

Since the substitution $y = t^4$ makes the differential equation true, $y = t^4$ is in fact a solution.

5. We know that at time $t = 0$, the value of y is 8. Since we are told that $dy/dt = 0.5y$, we know that at time $t = 0$

$$\frac{dy}{dt} = 0.5(8) = 4.$$

As t goes from 0 to 1, y will increase by 4, so at $t = 1$,

$$y = 8 + 4 = 12.$$

Likewise, we get that at $t = 1$,

$$\frac{dy}{dt} = .5(12) = 6$$

so that at $t = 2$,

$$y = 12 + 6 = 18.$$

At $t = 2$, $\dfrac{dy}{dt} = .5(18) = 9$ so that at $t = 3$, $y = 18 + 9 = 27$.

At $t = 3$, $\dfrac{dy}{dt} = .5(27) = 13.5$ so that at $t = 4$, $y = 27 + 13.5 = 40.5$.

Thus we get the following table

t	0	1	2	3	4
y	8	12	18	27	40.5

9. If $P = P_0 e^t$, then

$$\frac{dP}{dt} = \frac{d}{dt}(P_0 e^t) = P_0 e^t = P.$$

13. Since $dy/dx = -1$, the slope of the curve must be -1 at all points. Since the slope is constant, the solution curve must be a line with slope -1. Graph C is a possible solution curve for this differential equation.

17. Since $dy/dx = 2$, the slope of the solution curve will be 2 at all points. Any possible solution curve for this differential equation will be a line with slope 2. A possible solution curve for this differential equation is Graph F.

21. Since dy/dx is positive if y is positive, the slope of the solution curve is positive everywhere and increases as y increases, as in Graph A.

Solutions for Section 10.3

1.

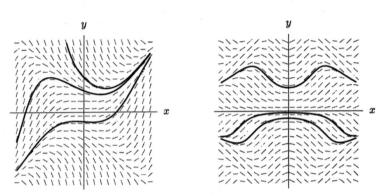

Figure 10.1

Other choices of solution curves are, of course, possible.

5. III. The slope field appears to be constant for a fixed value of y, regardless of the value of x. This feature says that y' does not depend on x, ruling out the formulas $y' = 1 + x$ and $y' = xy$. The differential equation $y' = 1 + y$ would have a slope field with zero slope at $y = -1$ and nowhere else, but the given slope field has two areas of zero slope, so $y' = 1 + y$ is ruled out and so is $y' = 2 - y$ for the same reason. This leaves $y' = (1 + y)(2 - y)$ as the correct answer, which fits the slope field as it has zero slopes at $y = 2$ and $y = -1$, positive slopes for $-1 < y < 2$ and negative slopes for $y < -1$ and $y > 2$.

9. (a) II   (b) VI   (c) IV   (d) I   (e) III   (f) V

13. As $x \to \infty$, y seems to oscillate within a certain range. The range will depend on the starting point, but the *size* of the range appears independent of the starting point.

Solutions for Section 10.4

1. The equation is in the form $dy/dx = ky$, so the general solution is the exponential function

$$y = Ce^{-0.14x}.$$

We find C using the initial condition that $y = 5.6$ when $x = 0$.

$$y = Ce^{-0.14x}$$
$$5.6 = Ce^0$$
$$C = 5.6.$$

The solution is

$$y = 5.6e^{-0.14x}.$$

5. Rewriting we get

$$\frac{dy}{dx} = -\frac{1}{3}y.$$

We know that the general solution to an equation in the form

$$\frac{dy}{dx} = ky$$

is

$$y = Ce^{kx}.$$

Thus in our case we get

$$y = Ce^{-\frac{1}{3}x}.$$

We are told that $y(0) = 10$ so we get

$$y(x) = Ce^{-\frac{1}{3}x}$$
$$y(0) = 10 = Ce^0$$
$$C = 10$$

Thus we get

$$y = 10e^{-\frac{1}{3}x}.$$

9. Since the rate of change is proportional to the amount present, we have $\frac{dQ}{dt} = kQ$. We know the constant of proportionality is $k = -0.0025$, so a differential equation for Q as a function of t is

$$\frac{dQ}{dt} = -0.0025Q.$$

The solution to this differential equation is

$$Q = Ce^{-0.0025t},$$

for some constant C. When $t = 20$, we have $Q = Ce^{-0.0025(20)} = C(0.951)$, so approximately 95% of the current ozone will still be here in 20 years. Approximately 5% will decay during this time.

13. **(a)** Since the growth rate of the tumor is proportional to its size, we should have

$$\frac{dS}{dt} = kS.$$

(b) We can solve this differential equation by separating variables and then integrating:

$$\int \frac{dS}{S} = \int k\,dt$$
$$\ln|S| = kt + B$$
$$S = Ce^{kt}.$$

(c) This information is enough to allow us to solve for C:

$$5 = Ce^{0t}$$
$$C = 5.$$

(d) Knowing that $C = 5$, this second piece of information allows us to solve for k:

$$8 = 5e^{3k}$$
$$k = \frac{1}{3}\ln\left(\frac{8}{5}\right) \approx 0.1567.$$

So the tumor's size is given by

$$S = 5e^{0.1567t}.$$

17. (a) Since we are told that the rate at which the quantity of the drug decreases is proportional to the amount of the drug left in the body, we know the differential equation modeling this situation is

$$\frac{dQ}{dt} = -kQ.$$

Since we are told that the quantity of the drug is decreasing, we include the negative sign.

(b) We know that the general solution to the differential equation

$$\frac{dQ}{dt} = -kQ$$

is

$$Q = Ce^{-kt}.$$

(c) We are told that the half life of the drug is 3.8 hours. This means that at $t = 3.8$ the amount of the drug in the body is half the amount that was in the body at $t = 0$, or in other words

$$0.5Q(0) = Q(3.8).$$

Solving this equation gives

$$0.5Ce^{-k(0)} = Ce^{-k(3.8)}$$
$$0.5C = Ce^{-k(3.8)}$$
$$0.5 = e^{-k(3.8)}$$
$$\ln(0.5) = -k(3.8)$$
$$k = \frac{-\ln(0.5)}{3.8}$$
$$\approx 0.182.$$

(d) From part (c) we know that the formula for Q is

$$Q = Ce^{-0.182t}.$$

We are told that initially there are 10 mg of the drug in the body. Thus at $t = 0$ we get

$$10 = Ce^{-0.182(0)}$$

$$C = 10.$$

Thus the formula is

$$Q(t) = 10e^{-0.182t}.$$

Substituting in $t = 12$ gives

$$Q(12) = 10e^{-0.182(12)}$$
$$= 10e^{-2.184}$$
$$Q(12) \approx 1.126 \text{ mg}$$

Solutions for Section 10.5

1. We know that the general solution to a differential equation of the form

$$\frac{dy}{dt} = k(y - A)$$

is

$$y = A + Ce^{kt}.$$

Thus in our case we get

$$y = 200 + Ce^{0.5t}.$$

We know that at $t = 0$ we have $y = 50$, so solving for C we get

$$y = 200 + Ce^{0.5t}$$
$$50 = 200 + Ce^{0.5(0)}$$
$$-150 = Ce^0$$
$$C = -150.$$

Thus we get

$$y = 200 - 150e^{0.5t}.$$

5. We know that the general solution to a differential equation of the form

$$\frac{dm}{dt} = k(m - A)$$

is

$$m = Ce^{kt} + A.$$

Factoring out a 0.1 on the left side we get

$$\frac{dm}{dt} = 0.1\left(m - \frac{-200}{0.1}\right) = 0.1(m - (-2000)).$$

Thus in our case we get

$$m = Ce^{0.1t} - 2000.$$

We know that at $t = 0$ we have $m = 1000$ so solving for C we get

$$m = Ce^{0.1t} - 2000$$
$$1000 = Ce^0 - 2000$$
$$3000 = Ce^0$$
$$C = 3000.$$

Thus we get

$$m = 3000e^{0.1t} - 2000.$$

9. In order to check that $y = A + Ce^{kt}$ is a solution to the differential equation

$$\frac{dy}{dt} = k(y - A),$$

we must show that the derivative of y with respect to t is equal to $k(y - A)$:

$$y = A + Ce^{kt}$$
$$\frac{dy}{dt} = 0 + (Ce^{kt})(k)$$
$$= kCe^{kt}$$
$$= k(Ce^{kt} + A - A)$$
$$= k\left((Ce^{kt} + A) - A\right)$$
$$= k(y - A)$$

13. (a) The quantity increases with time. As the quantity increases, the rate at which the drug is excreted also increases, and so the rate at which the drug builds up in the blood decreases; thus the graph of quantity against time is concave down. The quantity rises until the rate of excretion exactly balances the rate at which the drug is entering; at this quantity there is a horizontal asymptote.

(b) Theophylline enters at a constant rate of 43.2mg/hour and leaves at a rate of 0.082Q, so we have

$$\frac{dQ}{dt} = 43.2 - 0.082Q$$

(c) We know that the general solution to a differential equation of the form

$$\frac{dy}{dt} = k(y - A)$$

is

$$y = Ce^{kt} + A.$$

Thus in our case, since

$$\frac{dQ}{dt} = 43.2 - 0.082Q \approx -0.082(Q - 526.8),$$

we have

$$Q = 526.8 + Ce^{-0.082t}.$$

Since $Q = 0$ when $t = 0$, we can solve for C:

$$Q = 526.8 + Ce^{-0.082t}$$
$$0 = 526.8 + Ce^{0}$$
$$C = -526.8$$

The solution is

$$Q = 526.8 - 526.8e^{-0.082t}.$$

In the long run, the quantity in the body approaches 526.8 mg. See Figure 10.2.

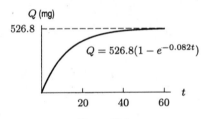

Figure 10.2

17. Let $D(t)$ be the quantity of dead leaves, in grams per square centimeter. Then $\frac{dD}{dt} = 3 - 0.75D$, where t is in years. We know that the general solution to a differential equation of the form

$$\frac{dD}{dt} = k(D - B)$$

is

$$D = Ae^{kt} + B.$$

Factoring out a -0.75 on the left side we get

$$\frac{dD}{dt} = -0.75\left(D - \frac{-3}{-0.75}\right) = -0.75(D - 4).$$

Thus in our case we get

$$D = Ae^{-0.75t} + 4.$$

If initially the ground is clear, the solution looks like the following graph:

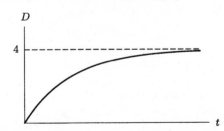

The equilibrium level is 4 grams per square centimeter, regardless of the initial condition.

21. (a) We know that the equilibrium solution is the solution satisfying the differential equation whose derivative is everywhere 0. Thus we must solve

$$\frac{dy}{dt} = 0.$$

Solving this gives

$$\frac{dy}{dt} = 0$$
$$0.5y - 250 = 0$$
$$y = 500$$

(b) We know that the general solution to a differential equation of the form

$$\frac{dy}{dt} = k(y - A)$$

is

$$y = A + Ce^{kt}.$$

To get our equation in this form we factor out a 0.5 to get

$$\frac{dy}{dt} = 0.5\left(y - \frac{250}{0.5}\right) = 0.5(y - 500).$$

Thus in our case we get

$$y = 500 + Ce^{0.5t}.$$

(c) The graphs of several solutions is shown in Figure 10.3.

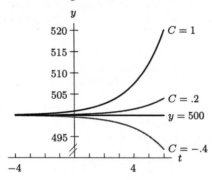

Figure 10.3

(d) Looking at Figure 10.3 we see that as $t \to \infty$, the value of y gets further and further away from the line $y = 500$. The equilibrium solution $y = 500$ is unstable.

25. (a) We know that the general solution to a differential equation of the form

$$\frac{dH}{dt} = -k(H - 200)$$

is

$$H = Ce^{-kt} + 200.$$

We know that at $t = 0$ we have $H = 20$ so solving for C we get

$$H = Ce^{-kt} + 200$$
$$20 = Ce^0 + 200$$
$$C = -180.$$

Thus we get

$$H = -180e^{-kt} + 200.$$

(b) Using part (a), we have $120 = 200 - 180e^{-k(30)}$. Solving for k, we have $e^{-30k} = \frac{-80}{-180}$, giving

$$k = \frac{\ln\frac{4}{9}}{-30} \approx 0.027.$$

Note that this k is correct if t is given in *minutes*. (If t is given in hours, $k = \frac{\ln\frac{4}{9}}{-\frac{1}{2}} \approx 1.62$.)

Solutions for Section 10.6

1. **(a)** If alone, the x population grows exponentially, since if $y = 0$ we have $dx/dt = 0.01x$. If alone, the y population decreases to 0 exponentially, since if $x = 0$ we have $dy/dt = -0.2y$.
   (b) This is a predator-prey relationship: interaction between populations x and y decreases the x population and increases the y population. The interaction of lions and gazelles might be modeled by these equations.

5. $\dfrac{dx}{dt} = x - xy, \quad \dfrac{dy}{dt} = y - xy$

9. This is an example of a predator-prey relationship. Normally, we would expect the worm population, in the absence of predators, to increase without bound. As the number of worms w increases, so would the rate of increase dw/dt; in other words, the relation $dw/dt = w$ might be a reasonable model for the worm population in the absence of predators.

   However, since there are predators (robins), dw/dt won't be that big. We must lessen dw/dt. It makes sense that the more interaction there is between robins and worms, the more slowly the worms are able to increase their numbers. Hence we lessen dw/dt by the amount wr to get $dw/dt = w - wr$. The term $-wr$ reflects the fact that more interactions between the species means slower reproduction for the worms.

   Similarly, we would expect the robin population to decrease in the absence of worms. We'd expect the population decrease at a rate related to the current population, making $dr/dt = -r$ a reasonable model for the robin population in absence of worms. The negative term reflects the fact that the greater the population of robins, the more quickly they are dying off. The wr term in $dr/dt = -r + wr$ reflects the fact that the more interactions between robins and worms, the greater the tendency for the robins to increase in population.

13. Sketching the trajectory through the point $(2, 2)$ on the slope field given shows that the maximum robin population is about 2500, and the minimum robin population is about 500. When the robin population is at its maximum, the worm population is about 1,000,000.

17. **(a)** Substituting $w = 2.2$ and $r = 1$ into the differential equations gives

$$\frac{dw}{dt} = 2.2 - (2.2)(1) = 0$$
$$\frac{dr}{dt} = -1 + 1(2.2) = 1.2.$$

   (b) Since the rate of change of w with time is 0,

$$\text{At } t = 0.1, \text{ we estimate } w = 2.2$$

   Since the rate of change of r is 1.2 thousand robins per unit time,

$$\text{At } t = 0.1, \text{ we estimate } r = 1 + 1.2(0.1) = 1.12 \approx 1.1.$$

   (c) We must recompute the derivatives. At $t = 0.1$, we have

$$\frac{dw}{dt} = 2.2 - 2.2(1.12) = -0.264$$
$$\frac{dr}{dt} = -1.12 + 1.12(2.2) = 1.344.$$

   Then at $t = 0.2$, we estimate

$$w = 2.2 - 0.264(0.1) = 2.1736 \approx 2.2$$
$$r = 1.12 + 1.344(0.1) = 1.2544 \approx 1.3$$

   Recomputing the derivatives at $t = 0.2$ gives

$$\frac{dw}{dt} = 2.1736 - 2.1736(1.2544) = -0.553$$
$$\frac{dr}{dt} = -1.2544 + 1.2544(2.1736) = 1.472$$

   Then at $t = 0.3$, we estimate

$$w = 2.1736 - 0.553(0.1) = 2.1183 \approx 2.1$$
$$r = 1.2544 + 1.472(0.1) = 1.4016 \approx 1.4.$$

21. (a) $\dfrac{dy}{dt} = \dfrac{0.6y - 0.8xy}{0.2x - 0.5xy}$; See Figure 10.4.

(b) $\dfrac{dy}{dx} = \dfrac{-y + 0.2xy}{-2x + 5xy}$; See Figure 10.5.

(c) $\dfrac{dy}{dx} = \dfrac{-1.6y + 2xy}{0.5x}$; See Figure 10.6.

(d) $\dfrac{dy}{dx} = \dfrac{-0.7y + 2.5xy}{0.3x - 1.2xy}$; See Figure 10.7.

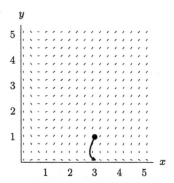

Figure 10.4: $\frac{dy}{dx} = \frac{0.6y - 0.8xy}{0.2x - 0.5xy}$

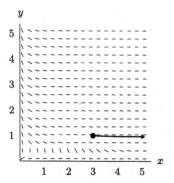

Figure 10.5: $\frac{dy}{dx} = \frac{-y + 0.2xy}{-2x + 5xy}$

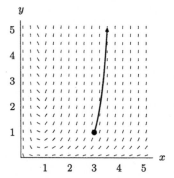

Figure 10.6: $\frac{dy}{dx} = \frac{-1.6y + 2xy}{0.5x}$

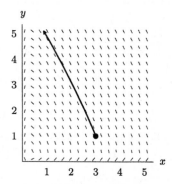

Figure 10.7: $\frac{dy}{dx} = \frac{-0.7y + 2.5xy}{0.3x - 1.2xy}$

Solutions for Section 10.7

1. Susceptible people are infected at a rate proportional to the product of S and I. As susceptible people become infected, S decreases at a rate of aSI and (since these same people are now infected) I increases at the same rate. At the same time, infected people are recovering at a rate proportional to the number infected, so I is decreasing at a rate of bI.

5. (a) $I_0 = 1$, $S_0 = 349$

(b) Since $\dfrac{dI}{dt} = 0.0026SI - 0.5I = 0.0026(349)(1) - 0.5(1) > 0$, so I is increasing. The number of infected people will increase, and the disease will spread. This is an epidemic.

9. The threshold value of S is the value at which I is a maximum. When I is a maximum,

$$\frac{dI}{dt} = 0.04SI - 0.2I = 0,$$

so

$$S = 0.2/0.04 = 5.$$

Solutions for Chapter 10 Review

1. Since $y = x^3$, we know that $y' = 3x^2$. Substituting $y = x^3$ and $y' = 3x^2$ into the differential equation we get

$$\text{Left-side} = xy' - 3y = x(3x^2) - 3(x^3) = 3x^3 - 3x^3 = 0.$$

Since the left and right sides are equal for all x, we see that $y = x^3$ is a solution.

5. We are told that y is a function of t (since the derivative is dy/dt) with derivative $2t$. We need to think of a function with derivative $2t$. Since $y = t^2$ has derivative $2t$, we see that $y = t^2$ is a solution to this differential equation. Since the function $y = t^2 + 1$ also has derivative $2t$, we see that $y = t^2 + 1$ is also a solution. In fact, $y = t^2 + C$ is a solution for any constant C. The general solution is

$$y = t^2 + C.$$

9. The general solution is

$$y = Ce^{5t}.$$

13. Multiplying both sides by Q gives

$$\frac{dQ}{dt} = 2Q, \quad \text{where } Q \neq 0.$$

We know that the general solution to an equation of the form

$$\frac{dQ}{dt} = kQ$$

is

$$Q = Ce^{kt}.$$

Thus in our case the solution is

$$Q = Ce^{2t},$$

where C is some constant, $C \neq 0$.

17. We know that the general solution to the differential equation

$$\frac{dH}{dt} = k(H - A)$$

is

$$H = Ce^{kt} + A.$$

Thus in our case we factor out 0.5 to get

$$\frac{dH}{dt} = 0.5 \left(H + \frac{10}{0.5} \right) = 0.5(H - (-20)).$$

Thus the general solution to our differential equation is

$$H = Ce^{0.5t} - 20,$$

where C is some constant.

21. We know that the general solution to the differential equation

$$\frac{dH}{dt} = k(H - A)$$

is

$$H = Ce^{kt} + A.$$

Thus in our case we factor out -0.5 to get

$$\frac{dH}{dt} = -0.5 \left(H - \frac{100}{-0.5} \right) = -0.5(H - (-200)).$$

Thus the general solution to our differential equation is

$$H = Ce^{-0.5t} - 200.$$

Solving for C with $H(0) = 40$ we get

$$H(t) = Ce^{-0.5t} - 200$$
$$40 = Ce^0 - 200$$
$$C = 240.$$

Thus the solution is

$$H = 240e^{-0.5t} - 200.$$

The graph of this function is shown in Figure 10.8

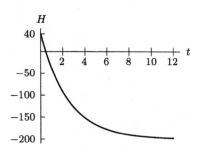

Figure 10.8

25. **(a)** If $C' = -kC$, and then $C = C_0 e^{-kt}$. Since the half-life is 5730 years, $\frac{1}{2}C_0 = C_0 e^{-5730k}$. Solving for k, we have $-5730k = \ln(1/2)$ so $k = \frac{-\ln(1/2)}{5730} \approx 0.000121$.

   (b) From the given information, we have $0.91 = e^{-kt}$, where t is the age of the shroud. Solving for t, we have $t = \frac{-\ln 0.91}{k} \approx 779.4$ years.

29. **(a)** $\dfrac{dy}{dt} = -k(y - a)$, where $k > 0$ and a are constants.

   (b) We know that the general solution to a differential equation of the form

$$\frac{dy}{dt} = -k(y - a)$$

   is

$$y = Ce^{-kt} + a.$$

   We can assume that right after the course is over (at $t = 0$) 100% of the material is remembered. Thus we get

$$y = Ce^{-kt} + a$$
$$1 = Ce^0 + a$$
$$C = 1 - a.$$

   Thus

$$y = (1 - a)e^{-kt} + a.$$

   (c) As $t \to \infty$, $e^{-kt} \to 0$, so $y \to a$.

   Thus, a represents the fraction of material which is remembered in the long run. The constant k tells us about the rate at which material is forgotten.

33. The closed trajectory represents populations which oscillate repeatedly.

Solutions to Problems on Separation of Variables

1. Separating variables gives

$$\int \frac{1}{P} dP = -\int 2dt,$$

so

$$\ln|P| = -2t + C.$$

Therefore

$$P = \pm e^{-2t+C} = Ae^{-2t}.$$

The initial value $P(0) = 1$ gives $1 = A$, so

$$P = e^{-2t}.$$

5. Separating variables gives

$$\int \frac{1}{u^2} du = \int \frac{1}{2} dt$$

or

$$-\frac{1}{u} = \frac{1}{2}t + C.$$

The initial condition gives $C = -1$ and so

$$u = \frac{1}{1 - (1/2)t}.$$

9. Separating variables gives

$$\frac{dz}{dt} = te^z$$

$$e^{-z} dz = tdt$$

$$\int e^{-z} dz = \int t\, dt,$$

so

$$-e^{-z} = \frac{t^2}{2} + C.$$

Since the solution passes through the origin, $z = 0$ when $t = 0$, we must have

$$-e^{-0} = \frac{0}{2} + C, \text{ so } C = -1.$$

Thus

$$-e^{-z} = \frac{t^2}{2} - 1,$$

or

$$z = -\ln\left(1 - \frac{t^2}{2}\right).$$

13. (a) Yes   (b) No   (c) Yes
  (d) No   (e) Yes   (f) Yes
  (g) No   (h) Yes   (i) No
  (j) Yes   (k) Yes   (l) No

17. Factoring and separating variables gives

$$\frac{dR}{dt} = a\left(R + \frac{b}{a}\right)$$

$$\int \frac{dR}{R + b/a} = \int a\,dt$$

$$\ln\left|R + \frac{b}{a}\right| = at + C$$

$$R = -\frac{b}{a} + Ae^{at}, \quad \text{where } A \text{ can be any constant.}$$

21. **(a)** The slope field for $dy/dx = xy$ is in Figure 10.9.

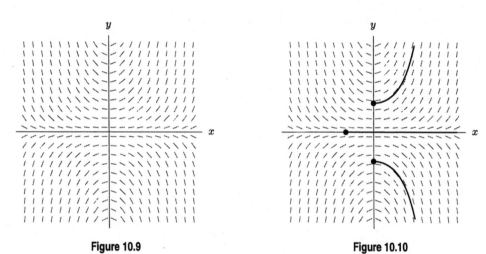

Figure 10.9   **Figure 10.10**

(b) Some solution curves are shown in Figure 10.10.
(c) Separating variables gives

$$\int \frac{1}{y}\,dy = \int x\,dx$$

or

$$\ln|y| = \frac{1}{2}x^2 + C.$$

Solving for y gives

$$y(x) = Ae^{x^2/2}$$

where $A = \pm e^C$. In addition, $y(x) = 0$ is a solution. So $y(x) = Ae^{x^2/2}$ is a solution for any A.

CHAPTER ELEVEN

Solutions for Section 11.1

1. Adding the terms, we see that

$$3 + 3 \cdot 2 + 3 \cdot 2^2 = 3 + 6 + 12 = 21.$$

We can also find the sum using the formula for a finite geometric series with $a = 3$, $r = 2$, and $n = 3$:

$$3 + 3 \cdot 2 + 3 \cdot 2^2 = \frac{3(1 - 2^3)}{1 - 2} = 3(8 - 1) = 21.$$

5. This is an infinite geometric series with $a = 1000$ and $r = 1.08$. Since $r > 1$, the series diverges and the sum does not exist.

9. We use the formula for the sum of a finite geometric series with $a = 1$, $r = 1/2$, and $n = 9$. We have

$$\text{Sum} = \frac{1 - (0.5)^9}{1 - 0.5} = 1.9961.$$

13. Each term in this series is half the preceding term, so this is an infinite geometric series with $a = 200$ and $r = 0.5$. Since $-1 < r < 1$, the series converges and we have

$$\text{Sum} = \frac{200}{1 - 0.5} = 400.$$

17. **(a)** Notice that the 6th deposit is made 5 months after the first deposit, so the first deposit has grown to $500(1.005)^5$ at that time. The balance in the account right after the 6th deposit is the sum

$$\text{Balance} = 500 + 500(1.005) + 500(1.005)^2 + \cdots + 500(1.005)^5.$$

We find the sum using the formula for a finite geometric series with $a = 500$, $r = 1.005$, and $n = 6$:

$$\text{Balance right after } 6^{\text{th}} \text{ deposit} = \frac{500(1 - (1.005)^6)}{1 - 1.005} = \$3037.75.$$

Since each deposit is $500, the balance in the account right before the 6th deposit is $3037.75 - 500 = \$2537.75$.

(b) Similarly, the 12th deposit is made 11 months after the first deposit, so the first deposit has grown to $500(1.005)^{11}$ at that time. The balance in the account right after the 12th deposit is the sum

$$\text{Balance} = 500 + 500(1.005) + 500(1.005)^2 + \cdots + 500(1.005)^{11}.$$

We find the sum using the formula for a finite geometric series with $a = 500$, $r = 1.005$, and $n = 12$:

$$\text{Balance right after } 12^{\text{th}} \text{ deposit} = \frac{500(1 - (1.005)^{12})}{1 - 1.005} = \$6167.78.$$

Since each deposit is $500, the balance in the account right before the 12th deposit is $6167.78 - 500 = \$5667.78$.

21. **(a)** The average amount in the body is $(65 + 15)/2 = 40$ mg.

(b) The average concentration for this patient (in milligrams of quinine per kilogram of body weight) is (40 mg)/(70 kg) = 0.57 mg/kg. This average concentration falls within the range that is both safe and effective.

(c) (i) Since this treatment produces an average of 40 mg of quinine in the body, a body weight W kg produces an average concentration below 0.4 mg/kg if

$$\frac{40}{W} < 0.4$$

so

$$W > 100.$$

The treatment is not effective for anyone weighing more than 100 kg (or about 220 pounds.)

(ii) A body weight W kg produces an average concentration above 3.0 mg/kg if

$$\frac{40}{W} > 3.0$$

so

$$W < 13.3.$$

The treatment is unsafe for anyone weighing less than 13.3 kg (or about 30 pounds.)

Solutions for Section 11.2

1. Right after the 5th deposit has been made, the 5th deposit has not yet earned any interest, the 4th deposit has earned interest for one year and is worth $2000(e^{0.06})$, the 3rd deposit has earned interest for two years and is worth $2000(e^{0.06})^2$, and so on. The first deposit has earned interest for 4 years and is worth $2000(e^{0.06})^4$. The total balance in the account right after the 5th deposit is

$$\text{Balance} = 2000 + 2000(e^{0.06}) + 2000(e^{0.06})^2 + 2000(e^{0.06})^3 + 2000(e^{0.06})^4.$$

This is a finite geometric series with $a = 2000$, $r = e^{0.06}$, and $n = 5$. Using the formula for the sum of a finite geometric series, we have

$$\text{Balance after the } 5^{\text{th}} \text{ deposit} = \frac{2000(1 - (e^{0.06})^5)}{1 - e^{0.06}} = \$11{,}315.60.$$

The balance right before the 5th deposit of \$2000 is $11{,}315.61 - 2000 = \$9{,}315.60$.

5. (a) The amount that must be deposited now is the present value of the annuity to fund this scholarship. The present value of the award made one year from now is $10{,}000(1.06^{-1})$, and the present value of the award made n years from now is $10{,}000(1.06^{-1})^n$. Since the awards start immediately, the 20th award is made 19 years from now. We have

$$\text{Present value} = 10000 + 10000(1.06^{-1}) + 10000(1.06^{-1})^2 + \cdots + 10000(1.06^{-1})^{19}.$$

This is a finite geometric series with $a = 10{,}000$, $r = 1.06^{-1}$, and $n = 20$. Using the formula for the sum, we have

$$\text{Present value of the annuity} = \frac{10000(1 - (1.06^{-1})^{20})}{1 - 1.06^{-1}} = \$121{,}581.16.$$

The donor must deposit \$121,581.16 now in order to fund this scholarship for twenty years.

(b) If the awards are to continue indefinitely, the present value is the sum of infinitely many terms:

$$\text{Present value} = 10000 + 10000(1.06^{-1}) + 10000(1.06^{-1})^2 + 10000(1.06^{-1})^3 + \cdots.$$

This is an infinite geometric series with $a = 10000$ and $r = 1.06^{-1} = 0.9433962$. Since $-1 < r < 1$, this series converges to the sum:

$$\text{Present value of the annuity} = \frac{10000}{1 - 1.06^{-1}} = \$176{,}666.67.$$

The donor must deposit \$176,666.67 now in order to fund this scholarship forever. We can check that this present value makes sense. After the first payment, the amount in the account is \$166,666.67. This amount will fund the awards forever because the interest earned on the account in one year exactly matches the money paid out of the account to fund the scholarship award: $\$166{,}666.67(0.06) = \$10{,}000$.

9. You earn 1 cent the first day, 2 cents the second day, $2^2 = 4$ the third day, $2^3 = 8$ the fourth day, and so on. On the nth day, you earn 2^{n-1} cents. We have

$$\text{Total earnings for } n \text{ days} = 1 + 2 + 2^2 + 2^3 + \cdots + 2^{n-1}.$$

This is a finite geometric series with $a = 1$ and $r = 2$. We use the formula for the sum:

$$\text{Total earnings for } n \text{ days} = \frac{1 - 2^n}{1 - 2} = 2^n - 1.$$

(a) Using $n = 7$, we see that

$$\text{Total earnings for 7 days} = 2^7 - 1 = 127 = \$1.27.$$

(b) Using $n = 14$, we see that

$$\text{Total earnings for 14 days} = 2^{14} - 1 = 16383 = \$163.83.$$

(c) Using $n = 21$, we see that

$$\text{Total earnings for 21 days} = 2^{21} - 1 = 2097151 = \$20,971.51.$$

(d) Using $n = 28$, we see that

$$\text{Total earnings for 28 days} = 2^{28} - 1 = 268435455 = \$2,684,354.55.$$

13. (a) In any given year, the number of units manufactured is 1000, the number of units in use that were manufactured the previous year is $1000(0.80)$ (since 20% of them failed), the number of units that were manufactured two years ago is $1000(0.80)^2$, and so on. We have

$$\text{Total number of units in use} = 1000 + 1000(0.80) + 1000(0.80)^2 + 1000(0.80)^3 + \cdots.$$

This is an infinite geometric series with $a = 1000$ and $r = 0.80$. Since $-1 < r < 1$, this series converges to a finite sum. We have

$$\text{Total number of units in use} = \frac{a}{1 - r} = \frac{1000}{1 - 0.80} = 5000 \text{ units.}$$

This sum, 5000 units, is the market stabilization point for this product.

(b) The total number in use after $n = 5$ production cycles is

$$S_5 = 1000 + 1000(0.80) + 1000(0.80)^2 + 1000(0.80)^3 + 1000(0.80)^4$$
$$= \frac{1000(1 - (0.80)^5)}{1 - 0.80}$$
$$= 3362 \text{ units.}$$

Similarly, we find the other values in Table 11.1

Table 11.1 *Market's approach to the stabilization point*

n	5	10	15	20
S_n	3362	4463	4824	4942

Solutions for Section 11.3

1. In 2003, the total quantity of oil consumed was 28.5 billion barrels; each subsequent year, the quantity is multiplied by 1.021. Thus, in 2004, we have $28.5(1.021) = 29.1$ billion barrels; in 2004 we have $28.5(1.021)^2 = 29.7$ billion barrels; and so on.

Year	2003	2004	2005	2006	2007	2008	2009	2010	2011	2012
Oil	28.5	29.1	29.7	30.3	31.0	31.6	32.3	33.0	33.7	34.4

In 2012, the total quantity of oil consumed is $28.5(1.021)^9$. Thus, we sum the geometric series:

$$\begin{aligned}
\text{Total consumption over decade} &= 28.5 + 28.5(1.021) + 28.5(1.021)^2 + \cdots + 28.5(1.021)^9 \\
&= \frac{28.5(1 - (1.021)^{10})}{1 - 1.021} = 313.498 \text{ billion barrels.}
\end{aligned}$$

5. Since the amount of ampicillin excreted during the time interval between tablets is 200 mg, we have

$$\text{Amount of ampicillin excreted} = \text{Original quantity} - \text{Final quantity}$$
$$200 = Q - (0.12)Q.$$

Solving for Q gives, as before,

$$Q = \frac{200}{1 - 0.12} \approx 227.27 \text{ mg}.$$

9. (a) We use the half-life to find the fraction of the drug remaining after one week. After a single dose of 100 mg of the drug, the quantity, Q, in the body decays exponentially so $Q = 100b^t$. We use the fact that the half-life is 18 weeks to solve for b:

$$50 = 100b^{18}$$
$$0.5 = b^{18}$$
$$b = (0.5)^{1/18}.$$

The fraction remaining after one week is $(0.5)^{1/18}$, or 0.9622238.

The steady state level right after a dose is the sum of an infinite geometric series with $a = 100$ and $r = (0.5)^{1/18}$. Since $-1 < r = (0.5)^{1/18} = 0.9622238 < 1$, the series converges and we use the formula for the sum of an infinite geometric series:

$$\text{Steady state level} = 100 + 100((0.5)^{1/18}) + 100((0.5)^{1/18})^2 + \cdots$$
$$= \frac{100}{1 - (0.5)^{1/18}}$$
$$= 2647.17 \text{ mg}.$$

(b) Since the steady state level is 2647.17 mg, the level of the drug in the body eventually passes 2000 mg. How many weeks does this take? The level of drug in the body after the nth dose (at the start of the nth week) is a finite geometric series with $a = 100$ and $r = (0.5)^{1/18}$. We have

$$\text{Amount after dose in } n^{\text{th}} \text{ week} = 100 + 100((0.5)^{1/18}) + 100((0.5)^{1/18})^2 + \cdots + 100((0.5)^{1/18})^{n-1}$$
$$= \frac{100(1 - ((0.5)^{1/18})^n)}{1 - (0.5)^{1/18}}.$$

We want to find the value of n for which this quantity is 2000. We simplify and use logarithms to solve for n.

$$2000 = \frac{100(1 - ((0.5)^{1/18})^n)}{1 - (0.5)^{1/18}}$$
$$20 = \frac{1 - ((0.5)^{1/18})^n}{1 - (0.5)^{1/18}}$$
$$0.755523 = 1 - ((0.5)^{1/18})^n$$
$$0.244477 = (0.5)^{n/18}$$
$$\ln(0.244477) = \frac{n}{18}\ln(0.5)$$
$$n = \frac{18\ln(0.244477)}{\ln(0.5)} = 36.58 \text{ weeks}.$$

The drug becomes effective in the 37th week.

13. Consumption of the mineral this year is 1500 kg, consumption next year is predicted to be $1500(1.04)$, consumption the following year is predicted to be $1500(1.04)^2$ and so on. Total consumption during the next n years is given by

$$\text{Total consumption for } n \text{ years} = 1500 + 1500(1.04) + 1500(1.04)^2 + \cdots + 1500(1.04)^{n-1}.$$

This is a finite geometric series with $a = 1500$ and $r = 1.04$. We have

$$\text{Total consumption for } n \text{ years} = \frac{1500(1 - (1.04)^n)}{1 - 1.04}.$$

We wish to find the value of n making total consumption equal to 120,000.

$$120,000 = \frac{1500(1 - (1.04)^n)}{1 - 1.04}$$

$$80 = \frac{1 - (1.04)^n}{1 - 1.04}$$

$$-3.2 = 1 - (1.04)^n$$

$$4.2 = (1.04)^n$$

$$\ln(4.2) = n\ln(1.04)$$

$$n = \frac{\ln(4.2)}{\ln(1.04)} = 36.59 \text{ years.}$$

Assuming the percent rate of increase remains constant, the reserves of this mineral will run out in 36 or 37 years.

17. If the mineral is used at a constant rate of 5000 m^3 per year, the total reserves of 350,000 m^3 will be used up in

$$\frac{350,000}{5000} = 70 \text{ years.}$$

Solutions for Chapter 11 Review

1. The formula for the sum of a finite geometric series with $a = 5$, $r = 3$, and $n = 13$ gives

$$\text{Sum} = 5 + 5 \cdot 3 + 5 \cdot 3^2 + \cdots + 5 \cdot 3^{12} = \frac{5(1 - 3^{13})}{1 - 3} = 3,985,805.$$

5. This is an infinite geometric series with $a = 75$ and $r = 0.22$. Since $-1 < r < 1$, the series converges and the sum is given by:

$$\text{Sum} = 75 + 75(0.22) + 75(0.22)^2 + \cdots = \frac{75}{1 - 0.22} = 96.154.$$

9. **(a)** (i) On the night of December 31, 1999:

First deposit will have grown to $2(1.04)^7$ million dollars.
Second deposit will have grown to $2(1.04)^6$ million dollars.
$\cdots$

Most recent deposit (Jan.1, 1999) will have grown to $2(1.04)$ million dollars.

Thus

$$\text{Total amount} = 2(1.04)^7 + 2(1.04)^6 + \cdots + 2(1.04)$$
$$= 2(1.04)\underbrace{(1 + 1.04 + \cdots + (1.04)^6)}_{\text{finite geometric series}}$$
$$= 2(1.04)\left(\frac{1 - (1.04)^7}{1 - 1.04}\right)$$
$$= 16.43 \text{ million dollars.}$$

(ii) Notice that if 10 payments are made, there are 9 years between the first and the last. On the day of the last payment:

First deposit will have grown to $2(1.04)^9$ million dollars.
Second deposit will have grown to $2(1.04)^8$ million dollars.
$\cdots$

Last deposit will be 2 million dollars.

Therefore

$$\text{Total amount} = 2(1.04)^9 + 2(1.04)^8 + \cdots + 2$$
$$= 2\underbrace{(1 + 1.04 + (1.04)^2 + \cdots + (1.04)^9)}_{\text{finite geometric series}}$$
$$= 2\left(\frac{1 - (1.04)^{10}}{1 - 1.04}\right)$$
$$= 24.01 \text{ million dollars.}$$

(b) In part (a) (ii) we found the future value of the contract 9 years in the future. Thus

$$\text{Present Value} = \frac{24.01}{(1.04)^9} = 16.87 \text{ million dollars.}$$

Alternatively, we can calculate the present value of each of the payments separately:

$$\text{Present Value} = 2 + \frac{2}{1.04} + \frac{2}{(1.04)^2} + \cdots + \frac{2}{(1.04)^9}$$
$$= 2\left(\frac{1 - (1/1.04)^{10}}{1 - 1/1.04}\right) = 16.87 \text{ million dollars.}$$

Notice that the present value of the contract ($16.87 million) is considerably less than the face value of the contract, $20 million.

13. The amount of additional income generated directly by people spending their extra money is $100(0.8) = $80 million. This additional money in turn is spent, generating another $(\$100(0.8))(0.8) = \$100(0.8)^2$ million. This continues indefinitely, resulting in

$$\text{Total additional income} = 100(0.8) + 100(0.8)^2 + 100(0.8)^3 + \cdots = \frac{100(0.8)}{1 - 0.8} = \$400 \text{ million}$$

17. **(a)** Let h_n be the height of the nth bounce after the ball hits the floor for the nth time. Then from Figure 11.1,

$$h_0 = \text{height before first bounce } = 10 \text{ feet,}$$
$$h_1 = \text{height after first bounce } = 10\left(\frac{3}{4}\right) \text{ feet,}$$
$$h_2 = \text{height after second bounce } = 10\left(\frac{3}{4}\right)^2 \text{ feet.}$$

Generalizing gives

$$h_n = 10\left(\frac{3}{4}\right)^n.$$

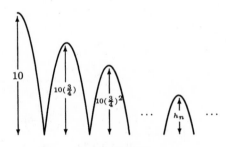

Figure 11.1

(b) When the ball hits the floor for the first time, the total distance it has traveled is just $D_1 = 10$ feet. (Notice that this is the same as $h_0 = 10$.) Then the ball bounces back to a height of $h_1 = 10\left(\frac{3}{4}\right)$, comes down and hits the floor for the second time. See Figure 11.1. The total distance it has traveled is

$$D_2 = h_0 + 2h_1 = 10 + 2 \cdot 10\left(\frac{3}{4}\right) = 25 \text{ feet.}$$

Then the ball bounces back to a height of $h_2 = 10 \left(\frac{3}{4}\right)^2$, comes down and hits the floor for the third time. It has traveled

$$D_3 = h_0 + 2h_1 + 2h_2 = 10 + 2 \cdot 10 \left(\frac{3}{4}\right) + 2 \cdot 10 \left(\frac{3}{4}\right)^2 = 25 + 2 \cdot 10 \left(\frac{3}{4}\right)^2 = 36.25 \text{ feet.}$$

Similarly,

$$D_4 = h_0 + 2h_1 + 2h_2 + 2h_3$$
$$= 10 + 2 \cdot 10 \left(\frac{3}{4}\right) + 2 \cdot 10 \left(\frac{3}{4}\right)^2 + 2 \cdot 10 \left(\frac{3}{4}\right)^3$$
$$= 36.25 + 2 \cdot 10 \left(\frac{3}{4}\right)^3$$
$$\approx 44.69 \text{ feet.}$$

(c) When the ball hits the floor for the nth time, its last bounce was of height h_{n-1}. Thus, by the method used in part (b), we get

$$D_n = h_0 + 2h_1 + 2h_2 + 2h_3 + \cdots + 2h_{n-1}$$
$$= 10 + \underbrace{2 \cdot 10 \left(\frac{3}{4}\right) + 2 \cdot 10 \left(\frac{3}{4}\right)^2 + 2 \cdot 10 \left(\frac{3}{4}\right)^3 + \cdots + 2 \cdot 10 \left(\frac{3}{4}\right)^{n-1}}_{\text{finite geometric series}}$$
$$= 10 + 2 \cdot 10 \cdot \left(\frac{3}{4}\right) \left(1 + \left(\frac{3}{4}\right) + \left(\frac{3}{4}\right)^2 + \cdots + \left(\frac{3}{4}\right)^{n-2}\right)$$
$$= 10 + 15 \left(\frac{1 - \left(\frac{3}{4}\right)^{n-1}}{1 - \left(\frac{3}{4}\right)}\right)$$
$$= 10 + 60 \left(1 - \left(\frac{3}{4}\right)^{n-1}\right).$$